Make:
FABRIC AND FIBER
INVENTIONS

Make:

FABRIC AND FIBER
INVENTIONS

Sew, Knit, Print, and Electrify Your Own Designs to Wear, Use, and Play With

KATHY CECERI

Maker Media, Inc.
San Francisco

Printed in Canada.

Published by
Maker Media, Inc.
1700 Montgomery Street, Suite 240
San Francisco, CA 94111

Maker Media books may be purchased for educational, business, or sales promotional use. Online editions are also available for most titles (*safaribooksonline.com*). For more information, contact our corporate/institutional sales department: 800-998-9938 or *corporate@oreilly.com*.

Publisher: Roger Stewart
Editor: Patrick DiJusto
Copy Editor: Rebecca Rider, Happenstance Type-O-Rama
Proofreader: Elizabeth Welch, Happenstance Type-O-Rama
Interior Designer and Compositor: Maureen Forys, Happenstance Type-O-Rama
Cover Designer: Maureen Forys, Happenstance Type-O-Rama
Indexer: Valerie Perry, Happenstance Type-O-Rama
Sewing icon designed by Smashicons from Flaticon

October 2017: First Edition
Revision History for the First Edition

2017-10-02 First Release

See *oreilly.com/catalog/errata.csp?isbn=9781680452273* for release details.

978-1-68045-227-3

Safari® Books Online

Safari Books Online is an on-demand digital library that delivers expert content in both book and video form from the world's leading authors in technology and business. Technology professionals, software developers, web designers, and business and creative professionals use Safari Books Online as their primary resource for research, problem solving, learning, and certification training. Safari Books Online offers a range of plans and pricing for enterprise, government, education, and individuals. Members have access to thousands of books, training videos, and prepublication manuscripts in one fully searchable database from publishers like O'Reilly Media, Prentice Hall Professional, Addison-Wesley Professional, Microsoft Press, Sams, Que, Peachpit Press, Focal Press, Cisco Press, John Wiley & Sons, Syngress, Morgan Kaufmann, IBM Redbooks, Packt, Adobe Press, FT Press, Apress, Manning, New Riders, McGraw-Hill, Jones & Bartlett, Course Technology, and hundreds more. For more information about Safari Books Online, please visit us online.

How to Contact Us

Please address comments and questions to the publisher:

Maker Media, Inc.
1700 Montgomery Street, Suite 240
San Francisco, CA 94111

You can send comments and questions to us by email at *books@makermedia.com*.

Maker Media unites, inspires, informs, and entertains a growing community of resourceful people who undertake amazing projects in their backyards, basements, and garages. Maker Media celebrates your right to tweak, hack, and bend any technology to your will. The Maker Media audience continues to be a growing culture and community that believes in bettering ourselves, our environment, our educational system—our entire world. This is much more than an audience, it's a worldwide movement that Maker Media is leading. We call it the Maker Movement.

To learn more about Make: visit us at *makezine.com*. You can learn more about the company at the following websites:

Maker Media: *makermedia.com*
Maker Faire: *makerfaire.com*
Maker Shed: *makershed.com*
Maker Share: *makershare.com*

To my mom, Deb Gradner, who showed me how to make my first toothpick knit pin—and how to teach a righty if you're a lefty. (You face each other, like in a mirror—easy!)

CONTENTS

Acknowledgments   *ix*

Introduction   *xi*

1   PRINT AND PAINT ON FABRIC   **1**

Color Theory and How to Blend Your Own Colors   3

Project: Capillary Action Sun Prints   7

Project: Tie Dye T-Shirts   14

Project: Science Ombre T-Shirt   22

Project: Simple Silkscreening   25

2   HAND-ME-UP WEARABLES AND FRIENDS   **31**

Upcycling Wool Sweaters   33

Project: Felted Wool Sweater Upcycling   35

Project: No-Sew Felted Sweater Wearables   37

Project: Felted Sweater Mittens   42

Project: Felted Sweater Hat   44

T-Shirts Reborn   48

Project: No-Sew T-Shirt Tote Bag   50

Project: Fringed T-Shirt   54

Project: Ball of T-Shirt Yarn   56

Project: T-Shirt Yarn Knotted Headband   60

Rescuing Unmatched Socks   65

Project: Design Your Own Sock Creature   66

3   FABRIC AND FIBER FURNISHINGS   **73**

Project: Coil Basket   75

Weaving on Looms   80

Project: Cardboard Weaving Loom   82

Building with Fabric   93

Project: Wearable Shelter   95

The Math of Quilting   101

Project: A Foldable Quilted Checkerboard   102

4 NEEDLE ARTS   **113**

Quick Crochet How-To   118

Project: Crochet Hot Cocoa Mug/Roll-Up Scarf   125

Quick Knitting How-To   131

Project: Tiny Toothpick Knitting   137

Project: Spool Knitter   140

5 SOFT TECH   **149**

How to Make an Electrical Circuit   152

Quick Soft Circuit How-To   156

Project: Sew a Soft Circuit Tester   167

Project: Soft Sensors   173

Project: Woven Audio Speaker   189

Index   *197*

ACKNOWLEDGMENTS

Thanks to the following people for their help and inspiration:

- ☐ Hanna, Carmen, and Eamon Heneghan and their mom Susie
- ☐ Rebecca Angel Maxwell (and a hearty shout-out to her nieces, Ayla and Thea Goldman)
- ☐ Jackie Reeve
- ☐ Beth Levine
- ☐ Julie Lewis
- ☐ Becky Stern
- ☐ Harriett Riddell
- ☐ Jesse Seay
- ☐ Hannah Perner-Wilson
- ☐ Leah Buechley
- ☐ The teams at Maker Media and Happenstance Type-O-Rama who help bring these books to life

INTRODUCTION
HOW TO BE A FABRIC AND FIBER INVENTOR

The oldest sewing needles ever found date back more than 20,000 years. They were made of bone and were probably used to make animal skins into clothes. Since that time, fabric and fiber inventions have changed the world we live in.

Like our early ancestors, we still spin the hair from sheep, llamas, and rabbits (not to mention fibers from silkworm cocoons) into yarn. We use old-fashioned tools to weave plants like cotton and flax (for linen) into yards of material. But today, in the laboratory, we also create fabric from plastics like nylon, polyester, and acrylic that can stop bullets (Kevlar), keep rain off while letting our skin breathe (Gore-Tex), and hold our sneakers closed (Velcro).

Natural and human-made textiles are used to make every kind of wearable, from simple robes to the latest fashions, and from protective gear for firefighters to space suits used by astronauts. Very soon, you may be wearing clothing that generates enough electricity to power your phone from your movements or from the sun, or light-up jackets that trigger turn signals on your bicycle when you lean to the right or left.

Over thousands of years, we have gone from making heavy woolen tents to shelter us in the desert, and canvas sails to power ships across the sea, to lightweight fabrics to take camping that can fold up and fit in a pocket. And we use decorative fabric and fiber to make furnishings for the home and create spaces in which to live and work.

All these things are possible because of people who came up with new kinds of textiles and new ways to work with them. Whether you are an artist or designer trying to create something beautiful or interesting, or an engineer or scientist who wants to solve a problem or fill a need, the invention process is pretty much the same:

- ☐ **Choose your challenge.** Decide on your goal to begin to narrow down what you will work on.

- ☐ **Brainstorm ideas.** It doesn't matter how crazy they are—wild ideas help you discover new angles to explore. Make sketches and write down notes. Then go through your ideas and pick one that looks promising.

- ☐ **Do your research.** What do you need to know about the challenge you're taking on? What have other people done that you can learn from? Look up news articles, collect images, check out videos, and talk to experts to help focus your ideas.

- ☐ **Create a prototype.** The best way to find out if your idea is a good one is to create a *prototype,* a quick model that lets you see what your invention will be like in real life. Making a prototype gives you a taste of what's involved in building your invention.

- ☐ **Test your idea.** Try out your invention to see if it does what you want it to do. Take notes, make measurements, and record your prototype in action. Get feedback from other people. Then go through the results to figure out what works and what doesn't.

- ☐ **Make improvements and keep testing.** Engineers use the word *iterate* to mean going through the steps of building and testing until you're happy with your creation.

- ☐ **Share your invention.** When you're ready to introduce your creation to the world, there are many ways to do it. Look for contests or exhibitions (like Maker Faire) where artists and engineers are invited to show off their work. You can post it online on a site like Instructables or Maker Share, sell it on Etsy, or crowd-fund it on Kickstarter or Indiegogo. You may even want to make your invention *open source* by publishing the instructions. That way, other people can try it out and create their own versions of your design.

- ☐ **Share your story.** If you really want to get people interested in your invention, talk about how and why you made it. Everyone loves a good story! And who knows, you may inspire them to become inventors themselves.

✂ FABRIC AND FIBER INVENTORS: BECKY STERN

Becky Stern (beckystern.com) is a rock star in the world of crafts and electronics. She has created video tutorials for *Make:* magazine and was the director of wearable electronics at Adafruit, a company that makes electronics for hobbyists and inventors. She is currently creating how-tos and online classes for the website Instructables.com on everything from textile arts (knitting, quilting, electronic embroidery) to jewelry making to cooking and hairstyles. Here she answers a few questions about her life and work:

FIGURE I-1: Becky Stern wearing one of her creations, a light-up choker. Credit: Andrew Tingle/Adafruit

You do a wide range of fabric and fiber crafts. How did you get started? And when did you begin adding electronic elements to them?

My mom taught me to sew and knit as a youngster. I started hand sewing copies of my Beanie Babies and knitting long scarves. I got a sewing machine for my 13th birthday and through high school I sewed my own purses and book bags. I even helped my mom sew my sister's wedding dress. I started adding LED lights and other electronics to my sewing projects in college when I took a class on interactive toy design.

How do you combine fashion and function so they work together? Is one more important than the other?

I try never to let a decorative element get in the way of the function of a project. It's harder than it sounds.

You produce great how-to videos and tutorials. What is the secret to teaching someone a new skill?

Empathy! Walk a mile in someone else's shoes, and it's easier to anticipate where they might get stuck when following your instructions.

What advice do you have for beginners who want to get into fabric and fiber arts/engineering/electronics?

Reject limitations. I believe it's important to find ways to experiment with your curiosity no matter what resources you have. Don't tell yourself that you can't do it. You will kill creativity before it ever gets started. Find mentors and experts who can show you new things and maybe even share their materials and tools to help you get started.

What crafts or activities do you do for fun?

I love knitting because it's relaxing, meditative, and productive at the same time. The same goes for motorcycle riding!

In this book, you'll learn the basic skills you need to make your own clothing, accessories, and household goods like baskets, blankets, and more. You'll even get to create your own *e-textiles*—electronic circuits you build into clothing and toys. But most of all, you'll discover the fun of using old and new techniques to turn your ideas into your own fabric and fiber inventions.

To help you get started, here's an introduction to some of the basic skills and supplies you'll need. You will find them helpful for a wide range of projects in this book, as well as projects you do on your own.

THE FABRIC AND FIBER SUPPLY BASKET

Most of the supplies you need for the projects in this book are easy to find in the crafts, sewing, hardware, or office supply section of your favorite store or online retailer. Some places to find specialized items are also listed.

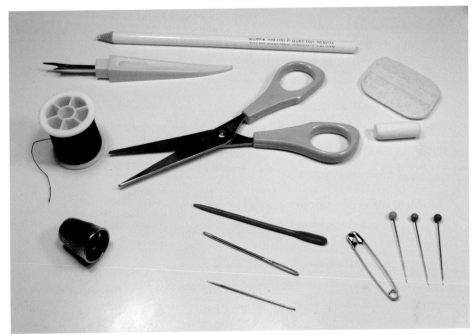

FIGURE I-2: Sewing equipment. Clockwise from left: thread, seam ripper, fabric pencil, fabric chalk, regular chalk, straight pins, safety pin, yarn needle (big and small), regular hand needle, thimble. Center: budget-priced sewing scissors.

Tools

Along with standard tools like pencils and rulers, you'll need some items used mainly for sewing or fiber arts.

Scissors

You will need scissors for almost every project in this book. Although ordinary scissors will work for almost all purposes, a sharp, clean pair of scissors will make your work easier. Serious crafters save special scissors that they use only on fabric, yarn, or thread so they stay nice and sharp.

Fabric Marker and Masking Tape

For many projects you will need to make temporary marks and lines on fabric. You can use a regular pencil or chalk and rub it away when you're done. Permanent marker will show up, but you can't erase it. You can, however, buy specialized fabric markers, pencils, or tailor's chalk that is made just for this purpose. For some projects, painter's blue tape or masking tape is used to mark straight lines.

Straight Pins/Safety Pins

Pins are used to hold pieces of fabric together before you sew them. If you're afraid of sticking yourself with a straight pin, safety pins are a good option (and recommended for quilting).

Needles

You may need needles for sewing thread, embroidery floss, and yarn, depending on the project. Sewing needles come in different lengths, and the hole for the thread (the eye) comes in different sizes. If you're not sure what size you may need, get a variety pack. A yarn or tapestry needle is meant to hold thick yarn and go through stiff fabric. It has a rounded point and comes in metal or plastic.

Needle Threader

When you have to make a thick yarn fit through the small eye of a needle, a needle threader comes in handy. Wire versions are cheap but don't usually last very long. A piece of thread or a narrow strip of paper will work just as well. To use a short piece of thread as a threader, just fold it into a loop. Holding the ends together, poke the loop through the eye of the needle. Insert the yarn through the loop. Then pull the ends of the thread back out through the eye, pulling the yarn with it.

To make a paper threader, take a short strip of paper that is narrow enough to fit through the hole of the needle. Fold it in half so the short ends meet. Poke the folded end through the hole of the needle, insert the yarn into the fold, and pull the threader and the yarn back through the eye of the needle, the same way you do with a piece of thread.

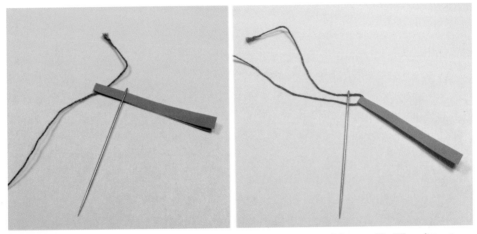

FIGURE I-3: **To use a needle threader, push it through the eye of the needle. Then insert the yarn through the threader (left image). Pull the threader back out of the eye, bringing the yarn with it (right image).**

Thimble

A thimble fits over your second finger to help you push a needle through a tough piece of material.

Seam Ripper

A seam ripper is useful for undoing machine sewing mistakes without damaging the fabric.

FIGURE I-4: **A thimble protects your fingers when you are pushing needles through fabric.**

Iron and Ironing Board

Use these for ironing out wrinkles, pressing creases where fabric gets folded, and layering fabric together with iron-on adhesive. Children should get adult help with using an iron.

Crochet Hook

The projects in this book were made with a US size H hook, which is equivalent to 5 mm.

Embroidery Hoop

Have on hand an inexpensive wood or plastic hoop in a medium to large size.

FIGURE I-5: **Embroidery hoop holding fabric**

Materials

This list includes suggestions for the best materials to use for the projects in this book, but in many cases you can get away with using scraps and leftovers. Check the remnant bin of your favorite fabric store for small quantities you can add to your collection.

Fabric

See project directions for what you'll need. These are the kinds of fabric used in the projects in this book:

- ☑ White cotton muslin (good for printing, dyeing, and silkscreening)
- ☑ Sheer (loosely woven and see-through) curtain material
- ☑ Ripstop nylon
- ☑ Lightweight cotton for quilting, by the yard (meter) or precut in strips or rectangles
  - ☉ Quilt roll—A bundle of precut strips 2.5 by 44 inches (roughly 6.4 by 112 cm) with colors or patterns that match
  - ☉ Quilt fat quarter—A bundle of precut pieces that are 18 by 22 inches (roughly 46 by 56 cm)
- ☑ Felt sheets
- ☑ Stiffened felt sheets—Extra-thick pieces may come in shapes that are easy to cut to size
- ☑ Aida cloth—Stiff cotton canvas with a loose weave, used for cross-stitching

Thread

Standard cotton blend thread is fine.

Yarn

Any kind can be used, but check directions for recommended weight (thickness) and fiber type. The varieties you are most likely to find in craft and hobby shops include acrylic, which is the least expensive, doesn't shrink if you wash it (because it's made out of plastic), and comes in many different weights and textures; cotton, which is thin, lightweight, and soft; and wool, which is the best quality, but is higher priced and can be itchy.

Notions and Fasteners

These items are nice to have but are not required.

- Snaps
- Buttons
- Hooks
- Velcro

Adhesives

Adhesives will help you connect pieces of fabric together without sewing them, and they can also give material more stiffness. These are some common adhesives:

- Iron-on/fusible strips and sheets (you can use fusible interfacing, which only has adhesive on one side, to stiffen a single piece of fabric)
- Fabric glue

Paints

Acrylic paint can be used for the projects in this book. A product called *textile medium* can be added to the paint to make it less stiff when it dries. You can also buy inks made specifically for silkscreening on fabric.

Specialized E-Textile Tools and Supplies

You will probably need to order some of these items online. In addition to Amazon.com or eBay, check out hobby electronic retailers, such as these:

- ☑ **www.sparkfun.com**
- ☑ **www.adafruit.com**
- ☑ **www.jameco.com**
- ☑ **https://lessemf.com**

Here is a list of some specialized tools and supplies you might need:

FIGURE I-6: Sewing an e-textile with conductive thread and LEDs

- ☑ Conductive thread—Sold in small bobbins (little plastic spools that fit into sewing machines) as well as larger cones or spools

- ☑ Variable resistor steel polyester yarn (The Cat. #306 at LessEMF.com [**https://lessemf.com**] will work.)

- ☑ LEDs with long wire leads, in sizes 3 mm, 5 mm, or 10 mm (gumdrop)

- ☑ Magnet wire (#30 or thinner—look for colors other than copper so you can see the insulated coating to make stripping easier.)

FIGURE I-7: A wire stripper

- ☑ Wire stripper—This is the best choice for removing insulation from wires. Scissors work in a pinch if you avoid cutting through the wire. Don't strip wire with your teeth!

Other E-Textile Tools and Supplies

You can find these items with other crafts, sewing, and jewelry-making supplies:

- ☐ Pliers—needle-nosed and round
- ☐ Wire cutter—look for a cutter with a small mouth, sometimes called a flush cutter, diagonal cutter, or nippers
- ☐ Soft jewelry wire or floral wire

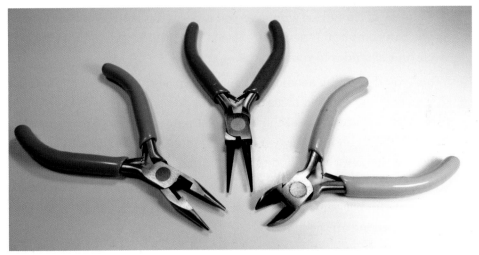

FIGURE I-8: **From left: needle-nosed pliers, round pliers, and a wire cutter**

Repurposed and Upcycled Materials

Thrift shops and garage sales (plus your own junk drawer or attic) are good places to find these items.

- ☐ Old wool sweaters
- ☐ Old sheer curtains
- ☐ T-shirts
- ☐ Musical greeting cards

IT'S SEW BASIC—TECHNIQUES TO KNOW

You need to know only a few sewing techniques to do the projects in the book.

Basic Hand Stitches

The following guidelines apply to most of the stitches you will do:

- Cut a piece of thread (or whatever you are sewing with) that is as long as you can comfortably sew with—usually a little longer than the length from your shoulder to your fingers.

- When you thread the needle, pull enough through the eye to allow you to hold onto the doubled piece of thread as you sew. This will keep the thread from slipping out of the needle. When you get close to the end of the thread (just enough for two or three more stitches), stop sewing. You will need the extra at the end to knot or anchor it.

- At the beginning and end of a line of stitching, make a knot in the end of the thread. For some projects, you will need to leave a "tail" of thread or yarn hanging off the fabric.

- Begin the first stitch by bringing the needle up from the wrong side of the fabric (the inside or the back of the piece you're working on). That way any knot or tail will be hidden from view.

- You may want to anchor the thread or yarn at the beginning or end of a line of stitching, or at a place where the stitching makes a sharp turn. This will hold your thread in place so it doesn't slip out and help reinforce corners. To anchor the thread, make two or three stitches in the same place. You can either make them right on top of one another, or criss-cross them.

- When you're sewing, make your stitches even. For beginning or young sewers, or when you're working with yarn or embroidery floss instead of thread, between ⅛ and ¼ inch (3 to 6 mm) is a good length.

- As you sew, pull your thread or yarn tight enough so it doesn't droop—but not tight enough that the fabric starts to pucker or bunch up.

To hide a tail of yarn or thread, you can sometimes weave it into your piece as you work. If it is still hanging loose when you're done, thread the tail on a needle and weave it inside other stitches. Trim the end so it doesn't show. Here are the stitches you need to know:

Running Stitch

The running stitch is a simple in-and-out stitch. Keep the stitches themselves as well as the spaces between them all the same length.

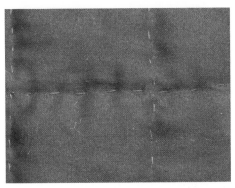

FIGURE I-9: **A running stitch**

Baste

When you baste fabric, the stitches go in and out like a running stitch, but they are longer and looser. Do not knot or anchor the ends. You will baste fabric when you want to pull the ends to gather the fabric into folds. You can also use it sometimes instead of pins to hold pieces of fabric together, or when you want to remove the stitches later.

FIGURE I-10: **Basting**

Backstitch

The backstitch is a nice secure stitch used to hold seams together. From the right side, it looks like a row of regular-sized stitches with no spaces in between. From the wrong side, it looks like double-sized overlapping stitches. The instructions here assume your regular-sized stitches are ¼ inch (6mm) long. If your regular stitches are longer or shorter, adjust the size to match.

1. Start by anchoring the first stitch by going over it two or three times. End with the needle on the wrong side of the fabric.

FIGURE I-11: **The backstitch**

2. Then bring the needle through from the wrong side to the right side, about ¼ inch (6 mm) away from the anchor stitches. In other words, leave a gap one regular stitch long between the stitch you just made and where you bring the needle up through the fabric. Go back across the gap and insert the needle back down through to the wrong side. Come up again about ¼ inch (6 mm) or one stitch length ahead of the stitch you just made and continue for as long as you need to.

Whip Stitch (also called Overcast Stitch)

A whip stitch is often used to finish a raw edge so it doesn't stretch or unravel. It is also good for holding trims or other materials, including wire, onto a piece of fabric.

Push the needle up from the wrong side to the right side of the fabric so that the tip pokes through next to the wire. Bring the needle over the wire and back down through the fabric. Make the next stitch the same way, close to the first stitch. Continue working along the rest of the wire the same way.

FIGURE I-12: **Whip-stitch around a wire.**

Slip Stitch (also called a Ladder Stitch or a Blind Stitch)

A slip stitch is a good way to join two pieces of fabric together from the right side of the project while also hiding the stitches. It's useful for finishing projects like pillows or stuffed toys and for holding folds and hems closed invisibly.

To join two pieces of fabric together along their folded edges, start by pushing the needle through one piece of fabric from the wrong side (inside the fold) to the right side. The needle should come out facing the gap between the two pieces of fabric, as close to the fold as possible. Bring the needle straight across the gap and through the other piece of fabric, from right side to wrong side (inside the fold). Take a small stitch and then bring the needle back through the first piece of fabric, close to the stitch before. Continue going back and forth. As you work, pull the thread or yarn tight enough so that the two pieces of fabric are touching and the stitches are hard to see.

FIGURE I-13: **The slip stitch**

Satin Stitch

The satin stitch is similar to a whip stitch, but done flat on a piece of fabric to create a smooth patch of stitches.

To start, bring the needle up to the right side of the fabric and back down on the other side of the area you want to cover. For the next stitch, bring the needle up right next to where you started the stitch before. Continue until the whole area is covered.

FIGURE I-14: **The satin stitch**

Basic Knots

Here are two ways to connect two strands of thread or yarn that are attached to your fabric:

Square Knot

Take the loose ends of the strands and bring the right-hand strand over the left-hand strand and up through the loop you have made. Repeat, but this time bring the left-hand strand over the right-hand strand. Pull tight.

FIGURE I-15: **The square knot is a double knot.**

Overhand Knot

Take the ends of both strands and loop them around and over both strands. Then bring the ends up through the center of the loop. Pull tight until the loop closes.

FIGURE I-16: **The overhand knot**

✂ USING A SEWING MACHINE

A basic sewing machine is all you need for the projects in this book. Every machine is different, so take a look at your machine's manual to see how it works. You should already know how to

- ☐ Thread the machine (from the spool at the top, through the thread guides, to the tension mechanism and the take-up lever, and through the eye of the needle at the bottom)

- ☐ Wind and insert a bobbin (the spool holding the lower thread)

- ☐ Raise and lower the presser foot (which holds the fabric down as you sew)

- ☐ Turn the hand wheel (to move the fabric along slowly)

- ☐ Use the foot pedal or knee control (to turn the motor on)

- ☐ Adjust the tension on the top and bottom thread so they are balanced

- ☐ Adjust the stitch length

FIGURE I-17: A vintage sewing machine

FIGURE I-18: Threading the needle. The presser foot is raised.

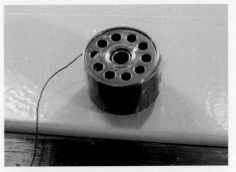

FIGURE I-19: A bobbin holding the lower thread

- Sew forward and backward (by changing the settings or turning the fabric around)

- Sew a straight seam following a guide (by lining the fabric up with markings on the throat plate or the side of the presser foot).

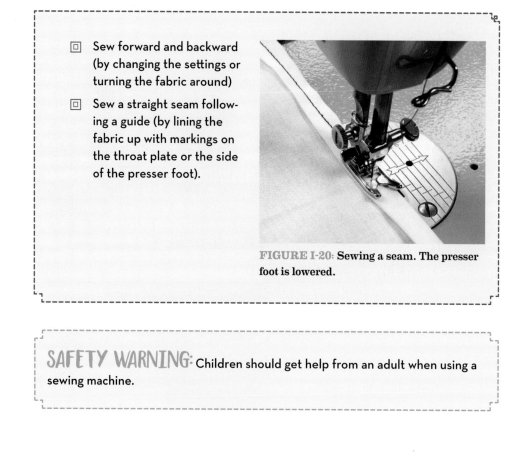

FIGURE I-20: Sewing a seam. The presser foot is lowered.

SAFETY WARNING: Children should get help from an adult when using a sewing machine.

PRINT AND PAINT ON FABRIC

Learn the secret science of working with colors and fabric.

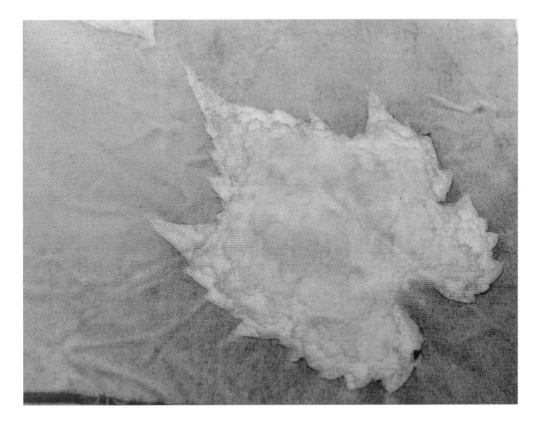

You may not think that figuring out how to add color to fabric is a big deal. After all, if you've ever splashed mud on your pants, or dribbled grape juice down your shirt, you've seen how some fibers soak up color and don't let it go. So it may surprise you to learn that the creation of paints, dyes, and other kinds of pigments played a major role in human history!

In early civilizations, only the most important people got to wear colorful clothes. That was one way other people could tell where you stood in society. Drab browns and dull yellows? You were probably a peasant who didn't have the time or the money to spruce up your wardrobe. Rich reds and vibrant blues? Your flashy clothing told the world that you were somebody rich and powerful.

The reason was that making colored dyes by hand was a long, hard process. It took so many workers to make enough dye for one piece of clothing that only the most powerful people in society could afford to wear them.

First, you had to gather enough plants, animals, or minerals to make your pigment. The deep shade of blue known as indigo—the same color used in blue jeans—came from the leaves of the indigo plant, which grows in India. The color crimson, a type of red, was created by grinding up a type of insect found inside oak trees in Greece. And to make one garment in "royal purple," which could be worn only by high priests and rulers, you first had to catch and boil 9,000 little snails found only in the Mediterranean Sea.

Then, once you had your pigment, you had to know what to add to create dyes that would stick to the fibers and not wash out or fade. The craftspeople who produced those brilliant colors kept their techniques a secret, which meant they were also important.

In the middle of the 1800s, all that changed—and fabric coloring again made history. The Industrial Revolution brought with it factories and machines that made it much cheaper to produce fabric and fiber. The demand for new clothes and other textiles exploded. Suddenly the old-fashioned way of producing dyes wasn't good enough.

Just like in ancient times, the color purple was particularly sought after and hard to create. The breakthrough came when a scientist accidentally found a way to make purple dye in the laboratory. That discovery inspired other scientists to develop even more colors. In fact, the search for synthetic dyes made in the laboratory gave birth to the era of modern chemistry. Some of the biggest chemical companies around today got their start thanks to the art of coloring fabrics.

In this chapter, you'll experiment with different ways to color fabrics with paint and dye. As you explore, see what new techniques and colors you can come up with!

COLOR THEORY AND HOW TO BLEND YOUR OWN COLORS

The projects in this chapter use lots of color. How can you tell which colors will go together well? That depends—it's a little bit of an art, and a little bit of a science.

Isaac Newton, who is famous for explaining how gravity works, also did research into light and color. He discovered that the white light of the sun is actually a combination of many colors. You can see those separate colors when light passes through a glass prism or water vapor in the air after a rainstorm and creates a rainbow. What's more, some colors can be created by combining other colors.

Today we know that light energy travels in a wave. Each color of the rainbow has its own wavelength. You may have noticed that the colors in a rainbow always appear in the same order, with red on top and violet (a shade of purple) on the bottom. That's because the light waves at the red end of the rainbow are longer and get smaller as you go toward the violet end. This arrangement of light waves in size order is called a *spectrum*. Human eyes can detect only a small range of wavelengths, which is called the *visible spectrum*.

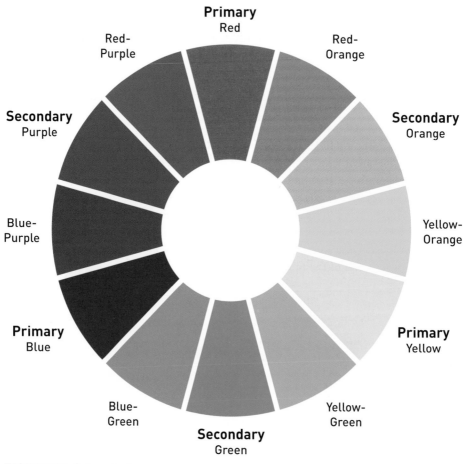

FIGURE 1-1: Color wheel

The colors you see when you look at paint work a little differently. Instead of going directly into your eye, light hits a spot of paint first and then bounces into your eye. But some of that light energy gets absorbed by the paint, so all you see are the wavelengths that remain. For instance, when white light falls on a red stop sign, almost all of the blue and green wavelengths are reflected back into your eye. (White paint reflects almost all the light energy that hits it. Black paint absorbs almost all the light that falls on it.) The result is that you need to mix different colors together depending on whether you're talking about light, paint, or dye or other transparent (see-through) pigments that let some light through.

Each type of coloring (light, paint, or dye) has three main colors, called *primary colors*, that can be used to make all the other colors. For light (such as the light of a TV screen), it is red, green, and blue. For paint (like the acrylic paint you will use in most of the projects in this chapter), the primary colors are red, blue, and yellow. For dyes, the primary colors are cyan (a type of blue), magenta (a type of red), and yellow.

In this chapter you'll be working with paints and dyes and seeing what happens when you mix colors together. To keep things simple, just remember that red, blue, and yellow (or shades of them) are the main colors to work with. With these colors, you can create pretty much all other colors.

You can use the color wheel (see Figure 1-1) to understand how different colors are related to each other. When you mix any two primary colors, you get a *secondary color*. For paints, these are green (blue + yellow), purple (blue + red), and orange (yellow + red). Mix a primary color with the secondary color next to it and you get an *intermediate color*, such as blue-green. And when you mix all three primary colors together, you get *tertiary colors*, which usually look brownish or grayish (for example, khaki green). Depending on the amount of each color you add, you can create different shades. For instance, aqua, turquoise, and teal are all combinations of green and blue. Look at the color wheel when you are deciding which colors will blend well!

✂ TIPS FOR WORKING WITH PAINTS AND DYES

You can get paints, dyes, and inks made specifically for the projects in this chapter, but when you're just starting out, it's fine to work with ordinary acrylic paint or to use dye kits made for beginners. Here is how to get the best results with whatever products you choose:

- ☐ Acrylic paint can be used for most of the projects in this chapter, but it may become stiff when dry. To soften it, thin the paint with water or a product called *textile medium*. Water may make the paint bleed through the fabric, so be sure to protect the work surface underneath.

- ☐ If you use acrylic paint, you need to heat it to "set" the paint so it doesn't wash out. The quickest way is to cover it with an old dish towel and iron it on high heat (no steam). You can also throw it in the dryer on high heat. The dryer may also help soften the painted area and make it more comfortable to wear.

- When working with fabric dyes, be sure to buy the correct dye for the fabric you are using. The easiest and safest dye to use comes ready-to-mix in squeeze bottles. All you do is add water, close the bottle and shake. Once it is mixed with water, you can use tie dye spray for any of the dye projects in this chapter.

- You'll get the best results if you machine or hand wash your fabric or garment before coloring it. This removes any protective chemicals that might interfere with the color. It also preshrinks the fabric. If the fabric needs to be dampened for your project, you can use it right out of the washing machine or wash basin.

- The projects in this chapter are designed for 100 percent cotton fabric. However, you may still get good results if your fabric is at least 50 percent cotton.

Here are some tips for minimizing the mess:

- To avoid getting color where you don't want it, cover the clothes you're wearing with an old oversized t-shirt, an apron, or an art smock. You can get disposable plastic aprons from art supply retailers.

- When working with dye, wear plastic or rubber gloves to avoid temporarily staining your skin.

- Also cover your work area with plastic sheeting, a disposable plastic tablecloth, or large garbage bags.

- If working indoors, you may want to cover the floor with a plastic drop cloth. If you can, work outside!

- Keep a roll of paper towels on hand for spills.

- Keep a bucket of water nearby and dip your gloved hands into it as you work to avoid spreading paint or dye onto anything you touch.

- Avoid rinsing dye in a porcelain sink, as it may stain. Stainless steel sinks are fine. If you are rinsing a dyed item onto the ground outdoors, pick a spot where the dye will not run directly into fresh water or onto a sidewalk or patio.

- When cleaning items that have been painted, printed, or dyed, wash them separately the first few times.

PROJECT:
CAPILLARY ACTION SUN PRINTS

⊙ MATERIALS

White or light-colored cotton cloth, washed

Acrylic paint in one or more colors

Water

(Optional) textile medium (for thinning regular acrylic paint) or transparent fabric paint

Objects to print (must block the light and lie flat on the fabric), such as these:

   Flat leaves and flowers

   String or yarn

   Buttons

   Dry pasta shapes

   Feathers

   Keys, combs, forks, paper clips, gears, nuts, rubber bands

   Cut-out shapes you buy or make yourself (plastic, wooden, metal, or craft foam—anything that is nonabsorbent)

   Kosher salt (the grains of salt soak up the liquid, creating interesting effects)

   Stiff screen or open plastic tray (to place on top of the other objects)

⊙ TOOLS

Plastic sheeting to cover your workspace (such as a plastic garbage bag, drop cloth, or disposable tablecloth)

(Optional) heavy cardboard sheet or wooden board to work on (cover with plastic sheeting)

Bowl or jar and disposable spoon for mixing paint

Foam brush

(Optional) spray bottle filled with water

Until digital cameras came along, most people took photographs on film and printed them out on paper. But one of the first types of photograph, known as sun prints, lets you make an image directly on paper—no film or camera needed! Early photographers like the British scientist Anna Atkins made sun prints to record plant specimens. Atkins took ferns, leaves, and flowers, lay them on the paper, and put them out in the sun. Special chemicals in the paper turned dark wherever the energy of the sun hit them. Areas that were covered stayed light. Artists still make sun prints—or to use the fancy name, heliographic art—to create special effects on paper or fabric.

A sun print is a type of *negative* image because the light and dark areas are reversed. Film cameras also produce negatives. When you press the button, the shutter opens and lets light in. Wherever the light hits the film, the film turns dark. To make a print, you have to reverse the process by shining light through the negative onto light-sensitive paper. In the final product, the light and dark areas are back to normal.

A true sun print is sometimes called a *cyanotype*, because the special ink used turns the paper

FIGURE 1-2: **This cyanotype sun print of a fern was made by the British botanist Anna Atkins in 1853 for a book about plants. Atkins was one of the first scientists to use photographs as illustrations. Image courtesy of the Open Content Program of The J. Paul Getty Museum.**

FIGURE 1-3: **A finished sunprint**

or fabric cyan, a type of blue. But you can get the look of a sun print by covering a piece of fabric with ordinary acrylic paint—and make it any color you like! In this version, the paint doesn't go through a chemical change. Instead, the wet paint is wicked away from areas that are wetter to areas that are drier. The sun dries the parts of the fabric that are uncovered faster than the parts that are covered. This leaves a lighter "shadow" behind that is shaped like the object you placed on the fabric, just like in a real sun print.

The wet paint moves because of *capillary action*, the ability of water to move inside materials containing a network of small openings. (A *capillary* is a very narrow tube.) It's the same thing that happens when you use a towel to wipe up a puddle of water on the kitchen counter. All substances, including water, are made up of tiny molecules. *Molecules* are the smallest part of a substance that still acts like that substance. Water molecules tend to stick together, which is why water forms drops and puddles with smooth rounded edges. Water molecules are also attracted to some other materials, like cotton cloth. When you blot a puddle with a cloth, the pull of the cotton is stronger than the force holding the water molecules together. The water gets pulled into the spaces between the threads in the cloth. These openings act like little straws that suck the water in. If the cloth is damp, it can pick up water even faster, because the water in the cloth also pulls on the water in the puddle.

Try making a Capillary Action Sun Print and see what kind of effects you can create!

1. Wet your cloth by first dipping it in a cup or bowl of water and then squeezing it out or by spraying it with water. It should be damp but not dripping. Spread the cloth out on the work board or table. Smooth out the cloth as much as possible. It should stick to the plastic, which makes it easier to work on.

FIGURE 1-4: **The damp cloth should stick to the plastic work surface.**

2. Put some paint in a bowl and mix it with an equal amount of water until it is very thin and runny. (For softer paint, use fabric paint, or add textile medium instead of water.) Dip the brush in the paint and cover the fabric well. For multiple colors, apply colors in separate areas and let the edges of each area run together. You can help them blend by spraying additional water where the colors touch.

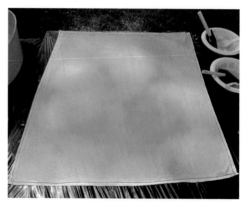

FIGURE 1-5: **Colors will overlap smoothly if you spray them with water.**

3. Arrange your objects on the fabric. Flatter objects that are touching the fabric will leave sharper outlines. The shadows of objects that are farther away will be lighter and fuzzier.

FIGURE 1-6: **Lay objects out any way you please.**

FIGURE 1-7: **The screen adds a soft background pattern and keeps objects from blowing away.**

4. Set your arrangement out in the sun to dry. Check on it every 15 minutes or half hour, depending on how hot and dry the air is. Gently peek under one of the objects to see if the fabric is still wet underneath. If it is, let it dry a little more and then check again.

FIGURE 1-8: **Peek under one object to see if the print is done.**

5. When the cloth is completely dry, remove the objects. You should see the shadows of the objects where the paint was wicked away.

6. If you used acrylic paint, iron your shadow print or throw it in a clothes dryer to set the paint. (See "Tips for Working with Paints and Dyes" at the beginning of the chapter.) Then display your print in a frame, or sew it onto a tote bag or throw pillow. Washing your print is not recommended.

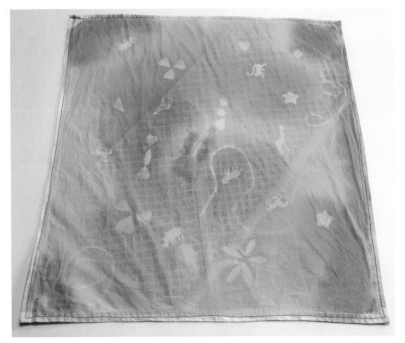

FIGURE 1-9: **The The end result. The red area is dye that ran from a feather.**

Extension:

You can turn a photograph into a Capillary Action Sun Print using a clear plastic negative. First, scan your image or open your photo file in a photo editing program and convert it to black and white. Then increase the contrast so there are strong areas of light and dark. Finally, invert (reverse) the dark and light areas to create a negative. Print out your image onto a sheet of plastic transparency film. (A copy shop can do this for you.) Tape the plastic sheet to the glass of a picture frame.

FIGURE 1-10: You can also use a photo with strong contrast to make a sun print.

Set the frame above the fabric on little "feet," to allow air to get in and water vapor from the evaporating paint to get out.

FIGURE 1-11: **A negative made from a digital photo and taped to a sheet of glass**

FIGURE 1-12: **Stacks of coins at each corner of the frame act like little feet to allow air to flow under it.**

FIGURE 1-13: **The finished product is a photo "print" made with paint on fabric.**

PROJECT:
TIE DYE T-SHIRTS

⊙ MATERIALS

1 or more bottles of spray tie dye, such as **Tulip One-Step Tie Dye**

(Optional) extra spray bottles for mixing your own colors

Spray bottle of water

Medium bucket (one or more)

Cotton t-shirt (kits can make several t-shirts or fabric projects)

(Optional) plastic or metal fork (for crafts only)

4–8 medium rubber bands for each item

Aluminum foil baking tray (or metal tray not used for cooking)

Metal cooling rack that fits in the metal tray

(Optional) metal tongs

Plastic bag big enough to hold dyed shirt and string or a twist-tie to seal the bag closed if the directions on your dye's packaging include them

---

Tie dye is a method of coloring fabric that creates patterns by covering some areas so the dye doesn't reach them. You bunch up or fold the fabric, "tie" it by wrapping rubbers bands around it, and then apply the color. The parts of the fabric inside the folds and wrinkles don't get as much dye as the outside. When you open the fabric back up, the lighter or undyed areas appear as designs.

Variations of tie dyeing are found in many parts of the world. The Japanese art of shibori involves wrapping fabric around poles, rocks, or other objects to create designs where the undyed fabric shows through.

FIGURE 1-14: **A spiral-dyed t-shirt**

SAFETY WARNING:

- ☑ Avoid handling powdered dye. It can be dangerous to breathe in and can cause an allergic reaction in some people if it gets on their skin. Look for kits where the powdered dye comes premeasured in bottles. Then, all you have to do is pour water in so you don't come in contact with the dry powder. You can also get refills in dissolvable pouches. If you must pour and measure powdered dye, wear a dust mask (like painters use).

- ☑ Always wear plastic or rubber gloves when working with dye. It can stain your skin and be hard to wash off.

- ☑ Don't use kitchen utensils that you intend to cook with again for dyeing! Find used or cheap versions at garage sales or the dollar store.

- ☑ Follow the "Tips for Working with Paints and Dyes" at the beginning of this chapter to keep the mess to a minimum.

Tie dying t-shirts first became popular in the 1960s because of their bright colors and cheery patterns. Back then, you had to dip the wrapped shirt in a bucket of dye, then retie it to dip it in another color. Today, spray bottle kits have made the process neater and easier. They let you put the color right where you want it to create more detailed patterns. Modern tie dye colors are also much brighter.

1. Protect your work area with plastic sheeting and cover your clothes with an apron or old t-shirt if needed. Dampen the t-shirt you're going to dye and then decide how you want to tie it. The easiest method is just to crumple it up and wrap several rubber bands around it. Here are a few other designs to try:

   ⊙ **Bull's-eye:** Pinch a spot on the shirt and pull it up to form a volcano-shaped hill. Wrap rubber bands around at one or more different heights. To turn it into a rose, push down the center of the "volcano" before tying it.

FIGURE 1-15: **For a bull's-eye, make a hill and wrap it with rubber bands.**

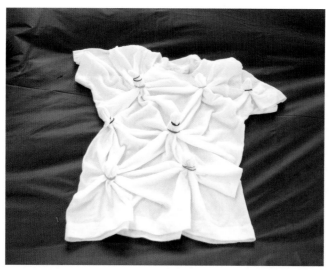

FIGURE 1-16: **Many small bull's-eyes**

⊙ **Stripes:** Fold the portion to be striped like a fan or accordion. Wrap the folded shirt with several rubber bands along its length. For a "V" shape, first fold the t-shirt in half lengthwise, laying the sleeves one on top of the other. Then fold the shirt at an angle, starting at one corner and going toward the opposite corner.

FIGURE 1-17: **For V-shaped stripes, fold the shirt in half.**

FIGURE 1-18: **Starting at one bottom corner of the shirt, fold it back and forth like an accordion or fan.**

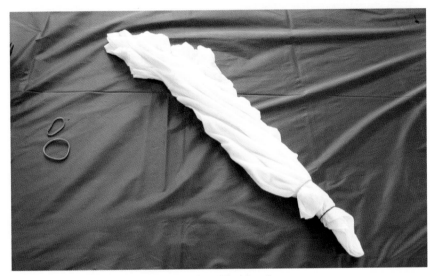

FIGURE 1-19: **Wrap the folded shirt in rubber bands.**

⊙ **Spiral**: Make a small ridge in the fabric where you want to center the spiral. Insert the raised fabric into the tines of a fork. Turn the fork like you are twirling spaghetti. Use your other hand to help the rest of the fabric curl around into the spiral. When you're done, the shirt should look like a flat disk. Wrap three rubber bands around the disk at regular intervals, dividing it up like slices of pie.

FIGURE 1-20: **Grab a ridge of fabric with a fork.**

FIGURE 1-21: **Twist the fork to curl the shirt up.**

FIGURE 1-22: **Shirt curled into a spiral**

2. Put on rubber gloves before you mix up the dye according to the directions on the packaging. If you want to combine dyes to make new colors, pour a little of the liquid dye into the extra bottles using the combinations listed in the Color Theory section at the beginning of this chapter.

FIGURE 1-23: **Wear gloves when handling dye.**

3. Place the metal rack in the metal tray. Lay the tied-up shirt on the rack. This will help keep the colors from mixing accidentally. Take a bottle of dye and squeeze it onto one section of the shirt marked off by rubber bands. If you want to leave little or no white showing, squeeze dye inside the folds as well. Squish the damp shirt around with your gloves to help the dye spread. Repeat with the other sections. Because the shirt is damp, the colors will run into each other, so choose colors that will mix well together. You can also overlap colors to mix them right on the shirt.

FIGURE 1-24: **Have extra squeeze bottles handy to blend new colors.**

4. Carefully flip the t-shirt over to add dye to the back where needed. Use the same colors in the same areas as you did on the front.

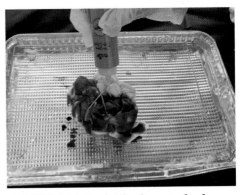

FIGURE 1-25: **Squirt the dye onto the damp shirt. Where colors touch, they will blend into new colors.**

5. Follow the directions on your dye's packaging for how to help the dye set. This will help keep it from running off the first time you wash your shirt. For Tulip One-Step Tie Dye, place your shirt in a plastic shopping bag and tie the bag shut tightly. Let it sit 6 to 8 hours or overnight.

FIGURE 1-26: **If your shirt needs to be wrapped in plastic, you can use a recycled shopping bag.**

FIGURE 1-27: **Tie the bag tightly.**

6. When you remove the shirt from the bag, rinse it in warm water. Remove the rubber bands and rinse it again until the water runs clear. Then wash the shirt by hand or machine in the hottest water possible and dry it.

7. It may take a few washes before the dye in your shirt doesn't run anymore, so wash it separately the first couple times. If the shirt is damp, don't put it on other clothes or a surface that it can stain.

FIGURE 1-28: **Rinse the shirt to remove excess dye.**

FIGURE 1-29: **Bull's-eye pattern after drying**

FIGURE 1-30: **V stripe pattern**

⊙ **MATERIALS**

T-shirt or other cotton fabric

At least 1 cup (250 ml) of liquid dye

Water

Bucket

Clothespins and string, or clothes hanger

(Optional) spray bottle of water

A good way to use up dye left over from tie-dying is to combine them to create an *ombre* effect—where the color appears in a range of brightnesses and tones, from darkest to lightest. This version adds a scientific twist by letting the fabric soak up the dye through capillary action. (See "Capillary Action Sun Prints" at the beginning of this chapter to learn how it works.) Since you are hanging the t-shirt vertically instead of letting it lie flat, the dye has to fight gravity to work its way up the fabric. That means the amount of dye in the shirt gets thinner the higher you go. Different pigments also climb at different rates, depending on the size of the molecules and how much each pigment is attracted to the fabric. If your color is a blend of different colors, or if you combine

FIGURE 1-31: **A blue-green ombre t-shirt after drying**

different colors of dye, you can watch them separate into different tones as they spread. Scientists use this process, which is called *chromatography*, to separate mixtures of chemicals by color. This project was adapted from an ombre project by artist Kathy Cano-Murillo (craftychica.com).

TIP: Science experiments can be tricky! You may have to try this one a few times to get the result you want. Make notes so you know what works and what doesn't the next time you do it. Here are some tips that may help:

- ☑ Darker colors may work better than lighter colors.
- ☑ Try using less water than usual to create a more concentrated mixture.
- ☑ To get a range of shades, try using a secondary color or mixing several leftover colors in the same color family together (see the Color Wheel in Figure 1-1).
- ☑ If your color isn't moving very fast, try letting it sit overnight.

---

1. Dampen your fabric with water. It should be moist, but not saturated, so that there's still room to absorb more liquid.

2. Set up your dyeing area. Protect the floor with a plastic sheet or do the project outside. Hang your t-shirt on a clothes hanger or on a string, clothesline style. Place a bucket underneath. Make sure the bottom of the shirt hangs into the bucket about 1 inch (2.5 cm) below the level of the dye.

3. Mix the dye according to the directions on the packaging. Pour it into the bucket. The liquid should be at least 2 inches

FIGURE 1-32: **Make sure the bottom of the shirt reaches below the level of the dye.**

(5 cm) deep. If necessary, add more water to make the liquid deep enough.

4. Let the shirt sit a few hours to get the effect you want. If the top of the shirt becomes too dry, spray it with a little water to keep it moist.

5. When you're happy with the ombre coloring, remove the shirt from the hanger. If the shirt bottom is very wet, use gloves to squeeze the excess dye out, working your way from lighter to darker.

6. Lay the shirt out flat on a sheet of plastic and let it dry, uncovered, for 6 to 8 hours or overnight. Follow the package directions for rinsing, drying, and washing the fabric.

FIGURE 1-33: **After a few hours, the colors in the dye have separated as they travel up the shirt.**

FIGURE 1-34: **Squeeze out the excess dye.**

PROJECT:
SIMPLE SILKSCREENING

◉ MATERIALS

Embroidery hoop, big enough to leave space all around your design

Adhesive vinyl (such as the peel-and-stick shelf liner popularly called "contact paper"), cut into a piece roughly the size of the embroidery hoop

Sheer fabric (loosely woven curtain fabric or traditional silkscreen silk), at least 3 inches (7.5 cm) bigger than the embroidery hoop on all sides

(Optional) blue painter's tape (or other wide tape)

Cotton fabric or t-shirt, as well as scrap fabric or paper for testing (blends that are at least 50 percent cotton may work)

(Optional) Sheet of cardboard (to go inside double-sided item such as a shirt or tote bag)

Thick acrylic paint or silkscreen ink (jar or tube)

◉ TOOLS

Pen or pencil

Plastic spoon (for scooping out paint from jar)

Squeegee (an old plastic gift card works fine)

(Optional) X-ACTO craft knife and cutting mat, or programmable vinyl cutter

---

Silkscreening is a way of stenciling an image onto fabric or other material. The paint or ink sits on the surface of the fabric, like a painting on a canvas. The result is a smooth, bright design that really pops.

Professional t-shirt silkscreening involves chemicals and big machinery. But you can get started with simple silkscreening using only basic craft supplies. This version

FIGURE 1-35: Silkscreened prints

was inspired by Sandra Roberts of Kaleidoscope Enrichment, who did color-changing silkscreen with young visitors at World Maker Faire New York in 2016 (**enrichscience.com/2016/10/color-changing-screen-printing**).

SAFETY WARNING: Younger fabric artists may need adult help cutting stencils out of vinyl. Use caution if using a craft knife to avoid cutting yourself or the furniture.

1. To start, trace the inside ring of the embroidery hoop onto the vinyl. This will be the outside edge of your stencil. If your hoop is much larger than your vinyl piece, you can extend the edges with painter's tape later on.
2. Draw your design on the vinyl. (If you draw on the paper backing, remember that the image will be reversed!) Keep the design simple. Avoid too many little details or very thin lines.
3. Cut the vinyl along the traced circle. Then cut out the areas that you want to print and set them aside. If you are using scissors, poke a hole in the middle of the shape with the tip. Cut a line from the hole to the outline of the shape, and then cut around the outline. Carefully remove the shape. The vinyl left around the hole(s) is your stencil, so avoid cutting or tearing it.

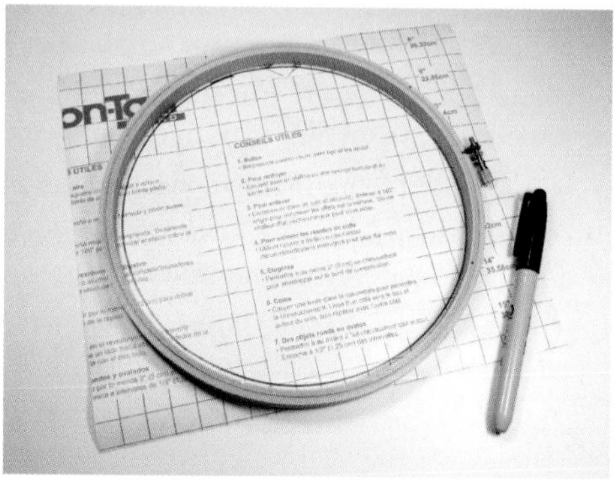

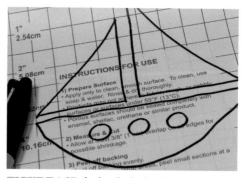

FIGURE 1-36: Trace your hoop onto your vinyl.

FIGURE 1-37: A simple design is easiest to cut.

4. Now it's time to make your screen. Cut a piece of fabric bigger than the embroidery hoop. There should be enough extra fabric hanging over the edge of the hoop to allow you to grab and stretch the fabric. Make it as tight as possible.

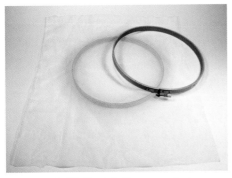

FIGURE 1-38: **Stretch the screen between the two hoops.**

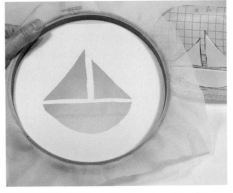

FIGURE 1-39: **Pull the screen tight.**

5. Trim any excess fabric that is in your way. Then turn your embroidery hoop silkscreen over so the fabric side is down. The inside of the hoop is the right side.

6. To transfer the vinyl to the screen, carefully remove the backing paper from your vinyl stencil. Press the stencil onto the right side of the fabric. Smooth out any wrinkles. If you have "islands" inside your main shape, like the portholes in the boat in Figure 1-37, cut them out of the excess vinyl you set aside. Attach them to the fabric the same way.

FIGURE 1-40: **Trim excess fabric.**

FIGURE 1-41: **Transfer the stencil to the screen.**

7. You're ready to print! If you are printing a t-shirt, tote bag, or other two-sided item, slip a sheet of cardboard inside to stop the ink from running through to the back. Lay your item to be printed (called a *substrate*) on a covered work surface. Lay the silkscreen, vinyl side up, on the substrate where you want your design to go.

NOTE: If you haven't done silkscreening before, make a few practice pieces on scrap fabric first.

8. Scoop or squeeze a small amount of paint along one edge of the silkscreen, away from the design. Hold the squeegee at an angle, tilted down toward the screen. Use it to drag the paint across the stencil, pushing color through the fabric where there are openings in the vinyl. Try to cover each section with one stroke. Make it as smooth and even as possible.

FIGURE 1-42: Spread some paint across the edge of the screen.

9. That's it! Carefully lift the screen to see how it looks. To copy the same design to multiple items, just move the screen to the next piece and keep printing. Let the printed pieces dry on a covered surface, or hang them to dry on a clothesline.

FIGURE 1-43: Slowly drag the squeegee across the opening in the stencil.

FIGURE 1-44: **Carefully lift the screen off the substrate.**

10. To save the stencil for another time, rinse the paint off the screen, being carefully not to shift the stencil around while it's wet. When the screen is dry, remove it from the hoop with the stencil still on it. Store the fabric and stencil flat.

Extensions:

☐ To print your design in a different color, rinse the paint off the screen and repeat with the new color.

☐ You can add details in different colors by breaking up your design into different layers. For example, to print a blue wave under an orange boat, make separate stencils for each color.

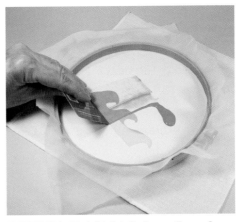

FIGURE 1-45: **Add details in another color.**

FIGURE 1-46: **A test piece with two colors**

HAND-ME-UP WEARABLES AND FRIENDS

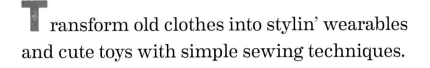

2

Transform old clothes into stylin' wearables and cute toys with simple sewing techniques.

When you take an existing object that has outlived its usefulness and turn it into something else, it's called *upcycling*. Upcycling is a great way to invent something. Instead of having to come up with an idea out of thin air, you can let the shape and texture of the material you are working with inspire you.

The tradition of turning used fabric items into new and useful household decorations and wearables goes way back. Worn-out pants, skirts, and shirts were cut up to make colorful rag rugs or warm patchwork quilts. Farm families took the cloth sacks they bought their flour or animal feed in and turned them into dresses, and in response, companies began printing attractive patterns on their bags to increase sales. In 1869, Swedish immigrant John Nelson patented the first sock knitting machines. He founded the Nelson Knitting Company in Rockford, Illinois, which became famous for its Rockford Red Heel Socks. The socks were seamless, so they were very comfortable. But the real reason they became so popular was that every pair came with directions for sewing them into sock monkeys!

In this chapter, you'll learn how to give old sweaters, t-shirts, and socks new life. If you don't have enough of the right kinds of old clothing on hand, keep an eye out when you visit thrift shops and garage sales, or trade unwanted clothing with your friends, like Swap-O-Rama-Rama inventor Wendy Tremayne.

✂ FABRIC AND FIBER INVENTORS: WENDY TREMAYNE

In 2005, writer and artist Wendy Tremayne started an event called Swap-O-Rama-Rama to help people learn to live well without buying a lot of stuff—including clothing. At a Swap-O-Rama-Rama, you donate old clothing that gets sorted into piles for anyone to use. Then you choose material to work with. Volunteer artists show you how to turn your chosen material into a new object or wearable using sewing machines, dye, silkscreen, and other techniques. Swap-O-Rama-Rama events have been held in over 150 cities around the world, including at Maker Faires in New York and the San Francisco Bay Area. Anyone can hold their own using the guide on Tremayne's website (swaporamarama.org/swap.htm).

NOTE: You can find directions for all the sewing stitches in this chapter in the "It's Sew Basic—Techniques to Know" section of the Introduction.

UPCYCLING WOOL SWEATERS

Have you ever accidentally washed a wool sweater in hot water? Chances are your sweater was a lot smaller when it came out of the dryer. That's because wool fiber—which is made from the fur of sheep, rabbits, alpaca, and other animals—tends to shrink when it gets washed.

There are two reasons for this. One has to do with the stuff that wool is made of. Just like human hair, fingernails, and even animal horns, wool is made of *keratin*. Keratin is a type of chemical known as a *protein*. When you look at a protein through a powerful microscope, you see that it comes in long chains that like to curl up. The keratin in wool can be stretched out straight like a spring that has lost its springiness. But if you heat the wool with hot water or hot air, the chemical chain absorbs the heat energy and springs back into its original shape. The tiny strands of keratin fold back up, and your wool sweater looks smaller.

The other reason wool shrinks is because each of the hairs that make up wool are covered in scales, sort of like the scales on a fish or a snake. All the scales face the way the hair grows. So when you comb the wool on a sheep, the hairs stay separate, just like when you comb your hair. But when you take a bunch of wool that has been cut off a sheep, the hairs face every which way. If you squeeze the wool, some of the scales catch on each other and become

FIGURE 2-1: **A sweater before and after felting**

tangled. As you continue to press on it, the wool gets packed together tighter and tighter. This process is called *felting*, and what you end up with is felt.

Felt is a great material to work with. Because the strands of fiber are locked together, you can cut it without worrying about the edges fraying. In this chapter you'll turn old wool clothing into felt by washing it in hot water. The combination of wetting it, heating it, and banging it around in your washing machine is enough to cause the wool yarn in an old sweater to shrink up and become a solid piece of felt.

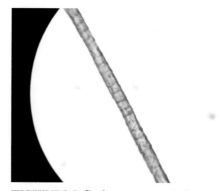

FIGURE 2-2: Scales on a strand of wool are visible under the microscope.

Different sweaters will produce different kinds of felt that can be used in many ways. Is it cushy and colorful? Try making mittens or a hat. Soft, thin, and flexible? Turn it into a scarf. And be sure to pay attention to the details: ribbing along the bottom and cuffs, fancy collars, decorative stitching, colorful patterns, or even buttons and zippers. You may be able to include them in your new pieces.

When you're looking for sweaters to repurpose, check the label to make sure they are between 75 and 100 percent wool. The label may say angora, mohair, or merino, which are all names for different types of wool. Some sweaters turn into felt better than others, so play around with a few to see what happens.

FIGURE 2-3: Individual stitches are visible on a sweater before shrinking.

FIGURE 2-4: After shrinking, the stitches all blend together.

PROJECT:
FELTED WOOL SWEATER UPCYCLING

FIGURE 2-5: Put sweaters into a net wash bag before washing them to keep lint from clogging the plumbing. The tennis ball helps the wool clump together.

Turn old sweaters into versatile felted wool with your washing machine and dryer!

1. Machine wash your sweater(s) in hot water. Set your machine for the longest available wash setting. You only need a small amount of detergent. To catch any wool lint that might clog up your washing machine or dryer, put the sweater in a net laundry bag or cotton pillowcase and tie it closed. Use the setting with the most agitation. If the sweater is colored, use a color catcher sheet (or two) to keep the dye from staining other clothes. You can also throw in some heavy pieces of clothing, towels, (clean) sneakers, or tennis balls to give the sweater something to rub against.

2. Dry your sweater on the hot setting. It is properly felted when you can no longer make out individual stitches. However, the real test is when you make a short snip into the fabric with a pair of scissors. Be careful to test your sweater in a spot where a cut won't matter. If the edge

stays straight and you don't get loose ends of yarn popping out, then it is ready to use as felt. Depending on the type of wool and how it is knitted, you may have to wash and dry it two or three times to get the right texture.

FIGURE 2-6: The clean edges on this cut show that the sweater has been properly felted.

Troubleshooting tips:

Some sweaters will never completely turn to felt or need extra attention. Here are some ways to deal with finicky knits:

- Wash the sweater by hand—or just the edges or other problem areas. Rub the parts you want to shrink on a bumpy surface, like a washboard, a bamboo sushi mat, or even a piece of bubble wrap. Rub the wool back and forth in the direction you want it to shrink.

- Stabilize an edge that looks iffy with stitching. A blanket stitch by hand with thread or yarn is decorative. Stitch straight along the edge by hand or with a machine. Or sew a machine zigzag stitch to make the edge into a ruffle.

- You can also just let the edge unravel and make the loose threads part of your design!

PROJECT:
NO-SEW FELTED SWEATER WEARABLES

◉ **MATERIALS**

1 old sweater, felted; size needed varies depending on the project

Chalk

◉ **TOOLS**

Ruler

Straight pins or paper clips

What to do with your felted sweaters? Here are some no-sew ideas that take just a few minutes.

Leg Warmers/Boot Toppers

1. Cut off both sleeves and pull them on your legs.

2. If you're going to wear the leg warmers inside or over the top of your boots, put your boots on now.

3. Decide how long you want your leg warmers to be. If they need to be shortened, mark the place(s) at which you want to cut on one leg with chalk.

FIGURE 2-7: A sweater before felting

FIGURE 2-8: Leg warmers made from the sweater in Figure 2-7. The remaining sweater is worn as a vest.

4. Take off the leg warmers and draw a straight line at the mark(s).

5. Cut along the line(s).

6. Measure the second sleeve against the first to make it match and repeat the cuts.

Arm Warmers

1. Pull one sleeve onto your arm so the end is even with your knuckles. If there is a seam, it should run up the inside of your arm. Adjust the sleeve if you like it a little bunched up.

2. If the sleeve is too long, take the chalk and mark where you want the top of the arm warmer to be. Also mark the place along the inside seam (see Figure 2-12) where the slit should go for your thumb to stick out.

3. Remove the sleeve and cut it where you marked it. Then try it on and adjust if needed.

4. When the first arm warmer is done, use it as a guide to mark and cut the second sleeve the same way.

FIGURE 2-9: **A sweater before felting**

FIGURE 2-10: **Arm warmers and a scarf made from the sweater in Figure 2-9**

FIGURE 2-11: **Cut the sleeves off the sweater.**

FIGURE 2-12: **Make marks to show where to cut.**

Scarf

1. Cut off the bottom hem of a sweater, or cut out a ring around the middle, and then cut it open at one of the side seams. (To turn it into an infinity scarf, after you've cut it open, twist one end so the back side is up and then stitch the ends back together.)

FIGURE 2-13: **After removing the sleeves, cut the side seams of the sweater open.**

2. To make a longer scarf, cut a U-shaped strip starting at the front hem, going up around the neckline, and then back to the front hem, as shown in Figures 2-13 and 2-14. Here's how you do it:

   A. Cut off the collar and the sleeves of the sweater.

   B. Cut each side seam open. Then cut the front of the sweater up the middle. The two front pieces are the ends of the scarf.

FIGURE 2-14: **Mark the upper edge of the scarf.**

C. Lay the sweater out flat. Trim the two front pieces so the edges from the front hem to the shoulder seams are straight.

D. Use the chalk to draw a line on the back connecting the outer edges of the shoulder seams. You can make the line straight, or follow the curve of the neckline.

Collar or Cowl

If the sweater has a nice collar, turtleneck, or other high neckline, cut it off and use it separately with other sweaters or tops.

FIGURE 2-15: **A sweater before felting**

FIGURE 2-16: **A cowl made from the sweater in Figure 2-15.**

Shrug (Short-Short Front-Opening Sweater)

A shrug can be as small as two sleeves connected with a strip across the back. You can also include (or add) a collar or a hood. If your felted sweater doesn't already open in the front, the first step is to cut open the front down the middle. Trim the two front pieces in whatever shape

FIGURE 2-17: **A shrug made from the sweater in Figure 2-15**

you like. If you want the front pieces to meet, you can add a button and buttonhole or loop to keep them attached.

FIGURE 2-18: **After removing the collar, cut the front of the sweater to the shape of the shrug.**

FIGURE 2-19: **The finished shrug**

PROJECT:
FELTED SWEATER MITTENS

⦿ **MATERIALS**

1 felted sweater body, large enough so you can trace around the fronts and backs of both your hands (sleeves may work for tiny hands)

Chalk

Yarn for sewing

⦿ **TOOLS**

Large-eyed needle

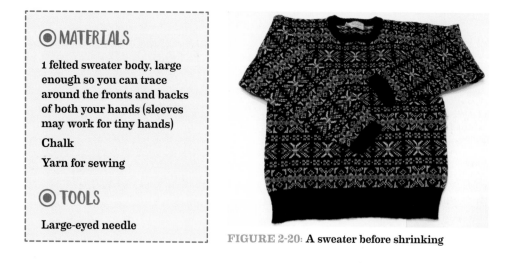

FIGURE 2-20: **A sweater before shrinking**

These mittens use the ribbing along the bottom of the sweater (or the sleeves, for tiny mittens) as the cuffs.

1. Place one of your hands on the body of the sweater (or the sleeve) so the top of the ribbing hits the top of your wrist. If you want to wear your mittens with the cuff rolled up, fold the bottom of the sweater up once and then put your hand on it. You'll have less sewing to do if you line up the pinky side of your hand along the side of the sweater at the seam. Hold your fingers together and your thumb out at an angle, away from the rest of your hand. Trace around your hand with the chalk, leaving space to sew a seam (about

FIGURE 2-21: **Mittens made from the sweater in Figure 2-20**

½ inch [1.25 cm]) all the way around. Repeat with your other hand, or cut out the first mitten shape and trace around that for the second.

FIGURE 2-22: **Line your hand up along the side and trace around it.**

FIGURE 2-23: **Cut out the mittens.**

2. Take the pieces for one mitten and make sure the cut edges match up. If you need to, hold them in place with straight pins or paper clips. Take some yarn and sew the edges together with a slip stitch. When you pull the stitches tight, they will almost vanish. Sew the other mitten the same way.

FIGURE 2-24: **To make a slip stitch, take small stitches in either side.**

FIGURE 2-25: **Pull the yarn tight to hide the stitches.**

PROJECT:
FELTED SWEATER HAT

⊙ **MATERIALS**

1 felted sweater body (or sleeve for a tiny hat)

Yarn for sewing

⊙ **TOOLS**

Large-eyed needle

Straight or safety pins

FIGURE 2-26: The pattern on this sweater gives the hat interest.

This hat is simple but looks great! It's basically a tube gathered at the top, which makes fitting it to your head extremely easy.

1. First, figure out how big your hat should be, and what area of the sweater you want to cut out to make it. If you use one of the bottom corners, the hem of the sweater can be the hem of the hat, and you only need to sew up one side seam. To find the length of the tube you will need, take a piece of yarn and use it to measure the distance from the top of your head to where you want the

FIGURE 2-27: A hat made from the sweater in Figure 2-26

bottom edge of the hat to sit. Add a little extra—about 2 inches (5 cm) and cut the yarn to this length. Then lay the yarn on the sweater, starting from one bottom corner (or wherever you want to start) and going up toward the shoulders. If you plan to wear the hat with the bottom rolled up, roll that part of the sweater up before you measure the height of the tube. Mark the length on the sweater with chalk.

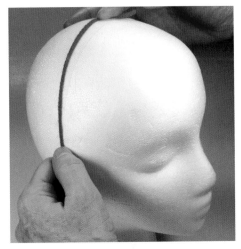

FIGURE 2-28: **Use a piece of yarn to measure the height of the hat on your head.**

FIGURE 2-29: **Use the yarn to show the distance on the sweater.**

2. (If you are using a sleeve, you can skip this step.) Next, find the circumference of your hat (the distance around the edge of the tube). Take another piece of yarn and measure around your forehead, right where you want the bottom of the hat to sit. Add 1 inch (2.5 cm) and cut the yarn to this length. Fold the yarn in half. Lay the folded yarn on the sweater along the bottom edge (or parallel to it). Mark this distance with the chalk. Cut out

FIGURE 2-30: **Use a piece of yarn to measure the distance around your head.**

the tube piece(s) along the lines you have marked. Use straight or safety pins to hold the cut edges together.

FIGURE 2-31: **Use the yarn to show the distance on the sweater.**

FIGURE 2-32: **Stitch up the side seam.**

3. Turn the hat inside out. Use a backstitch to sew the side(s) closed, about ½ inch (1 cm) in from the edge. Then baste around the top of the hat with big stitches, about 1 inch (2.5 cm) from the edge. Be sure to leave extra yarn hanging off at both ends. To gather the top, pull the ends of the yarn until the top edge of the tube is scrunched up as much as possible. Tie the ends together. You can wrap the ends around the bunched-up fabric and tie them once or twice more. Trim the ends of the yarn short. Spread the bunched-up fabric out to make it as flat as possible.

FIGURE 2-33: **Baste across the top of the hat.**

FIGURE 2-34: **Pull the basting yarn to gather the top of the hat.**

4. Turn the hat right-side out and try it on. If the gathered bunch of fabric inside the top is uncomfortable or looks lumpy, you can trim the extra fabric a bit.

Felted sweater extensions:

FIGURE 2-35: **The excess fabric at the top of the hat has been removed.**

☐ Mix and match pieces of sweaters—switch a collar from one sweater to another, or cut a sleeve into bands of different colors and textures. Add a wide short sleeve over a tight long sleeve to make multiple layers.

☐ Cut up small pieces to make flowers, bugs, or other 3D decorations. Sew them onto your other creations, or turn them into zipper pulls, pins, earrings, hair clips, and so on.

FIGURE 2-36: **The finished hat**

☐ Add belts, buckles, buttons, or other decorations.

☐ Sew conductive pads onto the fingertips of mittens, weave earphone speakers into a hat or ear warmer headband, or add LEDs, sensors, and other electronics. (See Chapter 5 for more detailed instructions.)

T-SHIRTS REBORN

If you ask people to name their favorite piece of clothing, chances are it will be a t-shirt. T-shirts are soft and comfy. They can be plain or colorful, and they can have words or pictures that tell the world about you. You can play in them, sleep in them, and in some places even work in them. They may not be fancy, but they're always fun to wear. And when you're done wearing them, they're great to upcycle!

Like wool sweaters, t-shirts are made from knitted fabric, which gives them their stretchiness. But instead of animal hair, t-shirts are usually made from cotton, a type of plant fiber. Sometimes it is combined with an artificial fiber like polyester, which is actually a kind of plastic. Because they are knit, t-shirts won't unravel easily when you cut them, which makes them easy to work with.

FIGURE 2-37: **T-shirt tote bag**

Even better, one type of t-shirt fabric, known as jersey, has a special property: when you cut it, the edges roll up. You can use this characteristic to add an interesting fringe to the unfinished bottom of a shirt. Or if you cut the t-shirt in long strips, you get soft, springy "yarn" that you can use just like any other kind of thick yarn or cord—for knitting, weaving, or knotting projects. (See the following chapters for project ideas.)

- ▢ To tell whether your t-shirt is made of jersey, look closely at the inside and outside of the shirt. The outside should look like chains of stitches going up and down in long columns. The inside should look like rows of loops going across. If it's hard to tell, test your t-shirt for curlability by cutting it in a place that won't show.

FIGURE 2-38: The outside of a jersey t-shirt (left) looks different than the inside (right).

- ▢ Even if it doesn't curl, you can still use any kind of t-shirt for the projects in this section, but you may not be able to fit beads on the fringe.

- ▢ If you have trouble getting a bead onto your fringe because the fabric is too thick, use a thin Phillips head screwdriver or other thin rounded rod to push the fabric through. Be careful not to break the bead.

FIGURE 2-39: Use a screwdriver or other sturdy rounded rod to push t-shirt fabric through the hole of a bead.

PROJECT:
NO-SEW T-SHIRT TOTE BAG

● MATERIALS

1 t-shirt (small to medium works best)

Painter's blue tape or masking tape

(Optional) pony beads or other large-holed beads

● TOOLS

Ruler

Pen

(Optional) chalk or straight pins

FIGURE 2-40: **A well-worn Maker Camp t-shirt**

Show off your favorite t-shirt design or saying by turning it into a tote bag. Stuff it in your pocket as an emergency shopping bag, or use it to store balls of yarn. If you need to carry small or thin objects like knitting needles, put them into another bag inside this tote, because there are small holes along the bottom. For added security, you can sew the bottom closed.

1. First turn the shoulders of the t-shirt into handles for your tote. To do this, cut off the sleeves of the t-shirt, including the seams between the sleeve and the body

FIGURE 2-41: **A colorful tote made from the t-shirt in Figure 2-40**

of the shirt. Then cut out the neckline. You can make this opening any shape, as long as you leave enough of the shoulder seam to make sturdy handles.

2. Next, figure out how deep you want to make the bag. The bottom of the bag will be formed by cutting the bottom of the t-shirt into strips that create a fringe. Then you will tie the strips of fringe in the front to the fringe in the back to hold the bottom closed. Hold your soon-to-be tote by the "handles" you created in Step 1 to see where you want the fringe to start. (Remember that the t-shirt fabric will stretch if the tote is carrying a lot of weight.) Mark the spot with chalk or pins. Stretch a piece of painter's or masking tape across the shirt so that the bottom of the tape runs along the line where you want to stop cutting the fringe.

3. Cut off the bottom hem from the shirt. Make sure that the cut bottom edge of the front and the back are even. Starting at one side (or inside the sewn side seam, if the shirt has them), cut strips of fringe about $\frac{3}{8}$ of an inch (1 cm) wide going from the bottom of the shirt to the bottom of the piece of tape. The strips

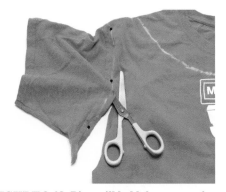

FIGURE 2-42: Pins will hold the seams of the sleeves together for neater cutting.

FIGURE 2-43: The neck and sleeves removed from the t-shirt

FIGURE 2-44: A line of tape will help you cut the fringe evenly.

don't need to be perfect, since they will roll inward. But if you need a guide, use the pen and the ruler to draw marks ⅜ of an inch (1 cm) apart along the piece of tape.

FIGURE 2-45: **Remove the hem of the shirt and put marks on the tape to show where to cut the fringe.**

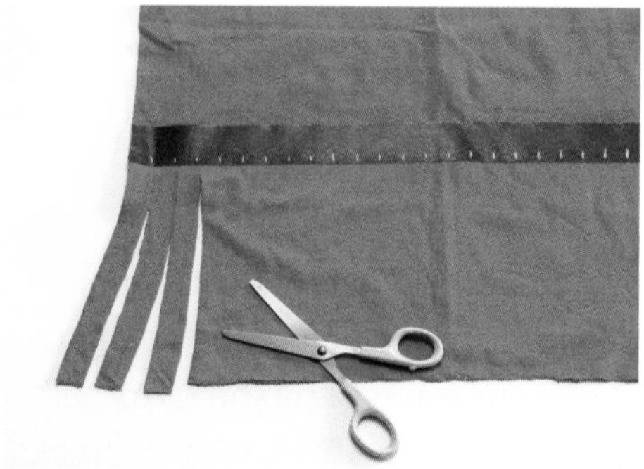

FIGURE 2-46: **Cut from the bottom edge to the bottom of the tape.**

4. Take each piece of fringe and stretch it down, away from the body of the shirt. If your shirt is jersey, the sides of the fringe will curl in, and the strip should become longer and thinner.

5. To close up the bottom of the tote, you will tie together each piece of fringe on the front of the shirt with the matching piece on the back. If you are not using beads, go on to Step 6. If you are using beads, now's the time to add them. Fold back every other pairing of fringe (the front piece and the back piece). Slip beads onto the remaining pairs of fringe. Tie all the pairs of fringe with beads as described in the Step 6. Then go back and tie all the remaining pairs.

FIGURE 2-47: **Stretch each piece of fringe to make it curl inward.**

FIGURE 2-48: **Flip every other piece of fringe up to get it out of the way.**

6. For each pair of fringe, take the piece in the front and tie it to the piece in the back. Use a square knot: tie right over left, then left over right. Pull the knot tight to make the gaps between the fringe pairs as small as possible. Repeat until you have tied all the fringe pairs across the bottom of the bag.

FIGURE 2-49: **Slide a bead on the front and back pieces of fringe, then knot the ends.**

FIGURE 2-50: **A square knot is like double-knotting your shoelaces.**

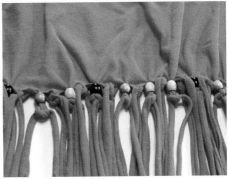

FIGURE 2-51: **When the beaded fringe is all tied, go back and tie the rest of the fringe.**

Extensions:

- For a large t-shirt, round the corners so the tote isn't too wide.

- If your fringe is long enough, use knots to make a design. (See the following Fringed T-Shirt project for directions.)

FIGURE 2-52: **Round off the corners of a wide t-shirt.**

PROJECT:
FRINGED T-SHIRT

⊙ MATERIALS

1 t-shirt that fits

Painter's blue tape or masking tape

(Optional) pony beads or other
large-holed beads

⊙ TOOLS

Ruler

Pen

(Optional) straight pins

(Optional) thin Phillips head screwdriver or other thin metal tool with
a rounded point

Get fancy with your fringe and create patterns by knotting it. For a real hippie vibe, use beads to emphasize the pattern.

1. Turn the shirt inside out and try it on. Figure out how high you want the fringe to go. Mark the spot with the pen, straight pins, or pieces of tape. Take the shirt off and lay it out flat (still inside out).

2. Follow Steps 3 and 4 from the No-Sew T-Shirt Tote Bag project.

FIGURE 2-53: **Fringe and beads makes a plain t-shirt pop!**

FIGURE 2-54: **Use tape to mark the top of the fringe.**

3. Starting with the front of the shirt, take the first two pieces of fringe on one side and tie the pieces together with a square knot (right over left, left over right) about 1 inch (2.5 cm) below the top. If you are using beads, add one before you tie the knot. Don't pull the knot so tight that the shirt bunches up—the pattern will look best if the shirt lies flat. Take the next two pieces and repeat the process until you have gone all the way around the bottom of the shirt.

4. For the next row, take one piece of fringe from one knot and one from the knot next to it. Tie these the same way, about 1 inch (2.5 cm) below the first row. Continue all the way around the shirt. Keep making knots the same way for as many rows as you like.

Extensions:

☐ Make the line that marks the top of the fringe at an angle, or in a V shape. The fringe will get shorter as the line gets closer to the bottom of the shirt.

☐ For fringed sleeves, trim off the hem of each sleeve and cut fringe almost to the shoulder seam. Pull and curl the same way as you did with the rest of the fringe.

FIGURE 2-55: **Cut the fringe from the bottom edge up to the bottom of the tape.**

FIGURE 2-56: **Tie each piece of fringe to the piece next to it along the front of the shirt, then continue along the back. For the next row, tie each piece of fringe to a piece from the pair next to it.**

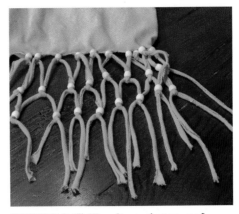

FIGURE 2-57: **The alternating rows of knots create a diamond pattern.**

PROJECT:
BALL OF T-SHIRT YARN

⊙ MATERIALS

1 t-shirt, the larger the better

Painter's blue tape or masking tape

(Optional) chalk

⊙ TOOLS

Ruler

Pen

(Optional) yardstick or long straight edge

FIGURE 2-58: T-shirt yarn in multiple colors

1. Cut off the bottom hem of your t-shirt. Also cut off the top of the shirt straight across from the bottom of one armhole to the other. Take the remaining torso piece and lay it out flat, making sure the bottom edges are even.

2. Attach a piece of painter's tape along one side of the t-shirt. If you need a guide for cutting strips of fabric evenly, use the ruler and pen to draw marks on the tape 1 inch (2.5 cm) apart, starting at the bottom. Then use chalk to make marks at the same height along the other side

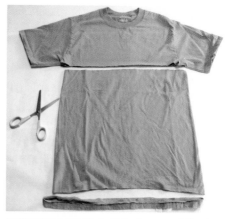

FIGURE 2-59: Cut off the hem and the top of the shirt.

of the shirt. Start to cut strips from the chalk marks on one side, stopping when you get to the tape. Repeat until you have cut the whole piece of shirt into strips.

FIGURE 2-60: **Lay tape along one side of the shirt and mark where you will cut.**

FIGURE 2-61: **Cut strips from the other side to the edge of the tape.**

3. Pick up the shirt and rotate it around so that the tape runs up the middle. Remove the tape. You are going to make cuts that connect each strip to the strip above it. Starting at the bottom-most cut, cut at an angle until you reach the next cut up. Continue connecting the cuts at an angle until all the strips have been connected. You should now have one long strip of fabric. If the ends are pointed, cut them straight across.

FIGURE 2-62: **The shirt completely cut (with the extra from the top removed)**

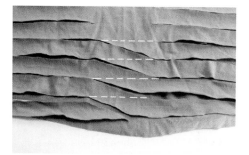

FIGURE 2-63: **Rearrange the shirt so the tape is in the middle.**

FIGURE 2-64: **Connect each strip to the strip above it with diagonal cuts.**

4. Take one end of the fabric strip and stretch it until the sides roll in and it becomes longer and thinner. Begin winding the stretched fabric into a ball. Stop as needed to stretch the next section as you wind. Continue until the entire strip is wound into the ball.

FIGURE 2-65: **Stretch the strips to make them curl inward.**

FIGURE 2-66: **Wind the yarn into a ball as you work.**

5. Save the yarn to use in any of the fiber arts projects in this book.

FIGURE 2-67: **Finished ball of t-shirt yarn**

✂ HOW TO CONNECT TWO PIECES OF T-SHIRT YARN

For projects that need more than one t-shirt's worth of yarn, or if you want to change colors in the middle of your project, you can easily connect a new strip onto an old one. Here's how:

1. Lay the end of the old strip flat in front of you. From the opposite direction, take the end of the new strip and place it on top, overlapping the old strip's end by about 1 inch (2.5 cm).

2. Squeeze the overlapping ends together and make a small vertical (up and down) cut. It will look like a buttonhole.

3. Take the other end of the new piece and push it through the cut from the back. Pull it tight. Stretch the pieces if necessary to get them to fit together smoothly.

FIGURE 2-68: Take the two pieces you want to connect and lay one end over the other. Then cut a small vertical slit through both strips.

FIGURE 2-69: Bring the other end of one strip through the slit.

FIGURE 2-70: Pull the end through.

FIGURE 2-71: When you pull the strip tight, the two strips will be securely connected.

PROJECT:
T-SHIRT YARN KNOTTED HEADBAND

FIGURE 2-72: **T-shirt yarn headband**

You only have to make short strips of yarn for this headband. The strips are held together using a type of a sailor's knot called the Carrick Bend.

1. Take a scrap piece of yarn and wrap it around your head like a headband. Mark the place where the ends overlap. This will be the final length of your headband.

FIGURE 2-73: **The body of the t-shirt used to make the headband**

2. Cut off the bottom hem of your t-shirt and lay it out flat, making sure the bottom edges are even.

Make a tiny mark along the side of the shirt about 1 inch (2.5 cm) wide above the bottom edge. If you need a guide, make another mark along the other side of the shirt and very lightly draw a line between them with a piece of chalk. Cut the strip. Repeat until you have six pieces of t-shirt yarn for each headband you are making.

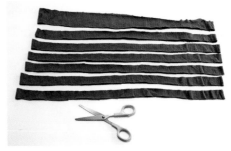

FIGURE 2-74: **Cut the shirt into six strips.**

3. Cut your pieces of t-shirt yarn open and stretch them until the sides roll in and the strips are longer and thinner. Lay the strips out flat. Make sure the strips are at least the length of your finished headband plus about 8 inches (20 cm).

> NOTE: If your shirt has sewn side seams, cut one of the seams right off. Try to position the remaining seam so it is at the back of the headband.

4. Divide the six pieces of t-shirt yarn into two strands with three pieces each. Take the first strand and twist it into a loop with the ends facing you.

5. Take the second strand and lay it over the ends of the loop in a large U. Bring the right leg of the loop over the U.

FIGURE 2-75: **Form one strand into a loop.**

FIGURE 2-76: Lay the second strand over it in a U shape.

FIGURE 2-77: Bring the right leg of the loop over the U.

6. Bring the upper left end of the U under the loop.

7. Take the upper right end of the U and thread it over-under-over the three strands at the top.

FIGURE 2-78: Bring the left end of the U under the loop.

FIGURE 2-79: Bring the right end of the U over-under-over.

8. Gently pull the knot tight. (It's easier if you have another person to help.)

FIGURE 2-80: **Pull the knot tight.**

9. Lay the headband out flat and smooth out the ends. Trim the ends so the entire piece is the length you measured for your headband.

FIGURE 2-81: **The finished knot**

10. Cut a rectangle out of t-shirt fabric about 4 inches (10 cm) by 5 inches (12 cm). Fold in about ¼ inch (7.5 mm) of the edges. (If you are using fabric or hot glue, glue them flat.) With the rectangle in front of you the long way, lay one end of the headband in the center of the rectangle. Bring the other end of the headband around so the ends are just touching. (If you are using glue, glue the ends of the headband to each other.) Wrap the top and bottom of the rectangle around the ends to make a tight tube. Glue the rectangle closed or sew around the sides of the rectangle to hold it closed and to hold the ends of the headband in place.

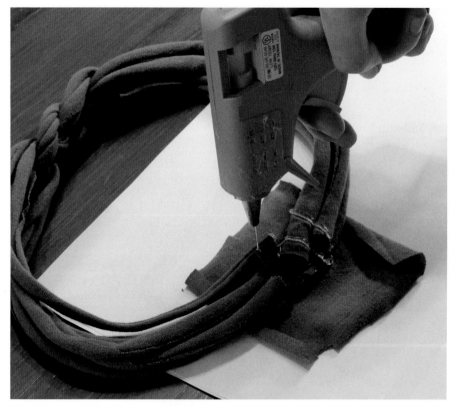

FIGURE 2-82: **Glue the ends of the knot together.**

Extension:

Create a wristband the same way, but make it thinner by using only four pieces of yarn.

RESCUING UNMATCHED SOCKS

Do you have a collection of socks that have lost their mates? Time to upcycle! Socks are knits just like sweaters and t-shirts, so they're easy to cut up and make into new items. For example, you can

- ☐ Trim a sock to fit over your water bottle to keep it cool.

- ☐ Fill the toe with dry rice, sew it closed, and use it as a hacky sack.

- ☐ Turn knee highs into arm warmers (see the directions using wool sweaters earlier in this chapter).

One of the most popular ways to repurpose old socks is to turn them into goofy creatures. Any kind of fiber will work. Go beyond the traditional sock monkey and invent a sock-based life form you can call your own!

FIGURE 2-83: **Colorful sock creatures**

PROJECT:
DESIGN YOUR OWN SOCK CREATURE

⦿ MATERIALS

1 sock (toddler or kid sizes are easier to work with; higher-cut socks give you more fabric to work with than low-cut anklets)

Needle and thread

Stuffing, such as polyester fiber fill (found in fabric stores) or dryer lint

(Optional) chopstick or other rounded stick

(Optional) buttons for eyes, other decorations

With a little sewing and stuffing it's easy to turn colorful socks into cuddly companions. By following a few basic design guidelines, you can make a lonely sock into any kind of cute animal, magical being, or weird creature you choose!

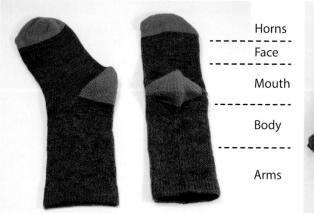

Horns
- - - - - - - -
Face
- - - - - - - -
Mouth
- - - - - - - -
Body
- - - - - - - -
Arms

FIGURE 2-84: **How the socks will be divided**

FIGURE 2-85: **A sock creature created from one of the socks in Figure 2-84**

1. First examine your sock and decide what kind of creature it wants to be. Draw a sketch of your idea, or let it evolve as you work on it. Here are some general guidelines:

   ⊙ **The heel of the sock is the face.** Add a couple of eyes to a sock with a heel that's a different color and you've got an instant face or mouth.

   ⊙ **The ankle of the sock is the body.** You can divide the ankle into two (or more) legs, or keep it as one piece to make a big-bellied bird or other creature. You can also use some of the material from the ankle to create additional parts, such as arms, a nose, ears, a tongue, or a tail. The edge of the cuff can be used as lips or an eyelid for a cyclops.

   ⊙ **The foot of the sock is the top of the head.** If you make the heel into a mouth, you have room on the foot of the sock for eyes and a nose. You can also cut the foot into hair, horns, or an antenna.

2. The next step is to cut up the sock. As an example, to transform a toddler sock into the creature in Figure 2-85, you would make three cuts:

   ⊙ Cut off the bottom of the ankle and save it to make arms.

   ⊙ Cut the remaining ankle section in half to make legs.

   ⊙ Cut the foot in half from the toe going toward the heel to make horns.

3. Turn the sock inside out. Take your needle and thread and use a running stitch to sew around the edges of the cuts you made in the head of the creature. In places where the stitches turn, like the space between the horns, go over the stitches again to make sure they don't pop open when you add the stuffing.

FIGURE 2-86: **Cut the toe in half for the horns.**

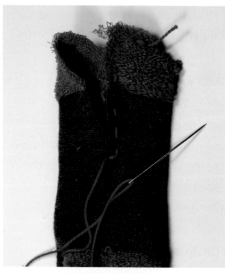

FIGURE 2-87: **Sew around the cut edges, taking an extra stitch at the end of the slit.**

FIGURE 2-88: **The finished stitching around the horns**

4. Sew the opening of the body closed—but make sure to leave an opening big enough to insert the stuffing. On the sock alien, you would leave one of the legs open.

5. Turn the sock right-side out again. Use your finger or a chopstick to push stuffing inside. You may have to tear off small pieces to fit into tight areas, like the horns. Fill the sock until you're happy with the puffiness. Then add stuffing right up to the opening.

6. Push in any stuffing that's hanging out of the opening, then sew it closed. Use a slip stitch if you don't want the stitches to show.

FIGURE 2-89: **Sew around the bottom cut edges, leaving an opening for the stuffing.**

7. To make the heel into a mouth, you need to pinch the heel to make a line running across it. Play around with the shape by poking and pinching it, then hold it in place with stitches. It can be short, long, straight or crooked, or curve into a smile. To begin, start at one end of the line you want to make and take a small stitch along the bottom of the heel. Then bring the needle up through the heel. The needle should come out through the bottom of the indent for the mouth. Then insert the needle into the top of the mouth and bring it out along the upper edge of

FIGURE 2-90: **The stuffed sock**

the heel. When you pull the thread tight, it should pinch the fabric of the heel together to start forming your line. If you need to, anchor the stitch at the top of the heel by going over it again. Start the next stitch about ¼ inch (6 mm) away, toward the center of the mouth line. Go back down through the mouth, the same way you did before, bringing the needle out at the bottom of the heel. Keep going up and down the same way until you have finished the line for the mouth.

FIGURE 2-91: **Start the mouth by taking a stitch below the heel of the sock.**

FIGURE 2-92: **Pinch the heel, and bring the needle out-in-out of the gap.**

8. Sew on buttons for eyes or any other decorations you want to add.

FIGURE 2-93: **Pull the thread up through the top of the heel, then take a stitch to hold it in place.**

FIGURE 2-94: **You can add eyes now or at the end.**

9. If you want to attach arms, cut the extra fabric you saved from the ankle of the sock in half. Fold each half so the cut edges are touching, then fold it in half again. The finished edge of the cuff will be the end of the arm. Use a slip stitch to sew the folded edges together and close up the cut bottom. Then attach the stitched end of each arm to the body, trying to hide the stitches inside the body or under the spot where the arm is attached.

FIGURE 2-95: **Use a strip of the cuff of the sock for the arm.**

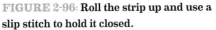

FIGURE 2-96: **Roll the strip up and use a slip stitch to hold it closed.**

FIGURE 2-97: **Sew the arm to the side of the Sock Creature.**

Extensions and adaptations:

- ☑ If you want your creature to bend so that it can sit, turn the sock around and use the heel as its behind.

- ☑ To make a creature with wild strands of hair like the little guy in Figure 2-83, follow these steps:

   1. Cut off the toe. (You can set it aside to use it later.)

   2. A little above the heel, go around the sock with a baste stitch.

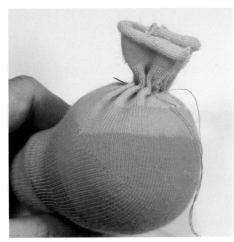

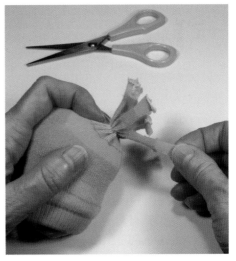

FIGURE 2-98: **Gather the top of the creature's head with stitches.**

FIGURE 2-99: **Cut the gathers into strips and pull to make strands of wild hair.**

3. Pull the ends of the thread together to gather the sock closed at that point. Take a few more stitches back and forth through the gathers to hold them in place.

4. Above the gathers, cut the remaining piece of the foot into eight strips, going from the top edge to just above the stitching. Stretch each strip until it rolls in on itself (with the right side of the sock material facing out).

☐ To make a no-sew creature, follow these steps:

1. Take an ankle sock and follow the preceding directions for making the wild strands of hair—but instead of gathering the sock with stitching, wrap a small rubber band around it.

2. Stuff the sock until the heel is rounded and the ankle is full.

3. Roll back the ankle cuff, and pull the toe you set aside over the opening. Tuck the edges of the toe inside the cuff, and then turn it back down.

4. If you need to, pull and push the stuffing until the creature stands on the edges of the cuff. Instead of buttons, draw on eyes with a permanent marker, or use peel-and-stick felt shapes.

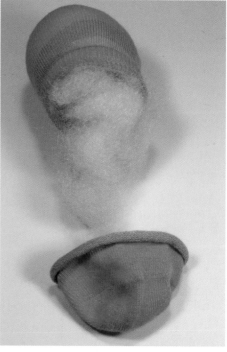

FIGURE 2-100: **Cut off the toe and stuff the creature as you wish.**

FIGURE 2-101: **Pull the toe over the bottom, and then roll down the cuff to hold it on.**

FABRIC AND FIBER FURNISHINGS

3

Create soft furnishings for your home—and even a soft shelter—using traditional fabric and fiber techniques.

Inventors often turn to soft materials like cloth and fiber when they want to make household containers and decorations, furniture, and even shelters that are portable and strong. The first humans to work with fiber may have been inspired by the nests of birds and other animals. They learned to weave grass, reeds, and vines into baskets so well-made that they could be used for cooking and even carrying water.

Around the world, nomadic people who traveled long distances in search of food made tents out of animal skin or fur so they could carry their homes with them. In tropical regions, people tied plant fibers into nets to catch prey, carry things, and even for hammock beds. Today, designers use high-tech fabrics to make lightweight backpacks, rugged camping gear, and emergency shelters.

In this chapter, you'll learn some techniques for turning fabric and fiber into objects that make life easier—and look great at the same time.

PROJECT:
COIL BASKET

⊙ MATERIALS

Core material such as

    Cotton rope

    Plastic clothesline

    Thick t-shirt yarn (see Chapter 2 for directions)

Wrapping material such as

    Ordinary yarn (can be doubled to make it thicker)

    Thin t-shirt yarn

    Torn strips of fabric about ½-inch (1.25-cm) wide

⊙ TOOLS

Yarn needle (sturdy plastic or metal needle with a rounded point and a large hole)

FIGURE 3-1: Fabric coil basket

Coiled baskets are quick and easy to make. A length of core material is curled into a spiral that supports the bottom and sides. A wrapping material is woven around the coils to connect them and hold them in place. Try experimenting with different sizes and materials, and see how many designs you can create.

1. Start with roughly 12 feet (4 m) of core and 3 feet (1 m) of wrapping. Thread one end of the wrapping through the needle for taking stitches. You will start wrapping with the other end.

2. Cut the end of the core at an angle. That will make it coil up with less lumpiness. Lay the free end of the wrapping over the end of the core in opposite directions. Let them overlap about an inch (2.5 cm).

FIGURE 3-2: **Lay the yarn along the core.**

3. Begin winding the wrapping tightly around the core for about 2 or 3 inches (5–7 cm), covering the bit of yarn at the same time.

4. Then coil the wrapped core into a spiral to start the bottom of the basket. Take the needle and poke it through the center of the loop and pull the yarn tight. Go around the coil and take a few more stitches into the center to hold it securely.

FIGURE 3-3: **Wrap the yarn around the core.**

5. Now wind the wrapping another four or five turns around the core (use fewer turns if the wrapping is thick). For the next stitch, insert the needle into the space between this coil and the one before it. Continue wrapping and stitching until you've

FIGURE 3-4: **Curl the rope into a spiral.**

got a flat circle big enough for the base of your basket. You may need to adjust the number of turns to avoid bumping into stitches from the inside rows.

6. To start a new piece of yarn, leave about 1 inch (2.5 cm) lying along the core. Lay the same length of the new yarn next to it. Continue wrapping from where you left off, covering both ends of the wrapping so that they don't show.

7. When the bottom is big enough, begin to shape the sides. Instead of placing the next round of core on the outside of the circle, put it on top. Wrap and stitch as before. As you go around, build the coils up to become the walls of your basket. You can shape the side straight up, or let it flare out or close back in.

8. If you need to, add a new piece of core to keep building the sides. To keep the coil smooth, cut the ends of the old piece and the new piece at matching angles. Fit them together as closely as possible. Wind the wrapping around the splice tightly to hold it in place.

FIGURE 3-5: **Continue wrapping and taking stitches into the first coil.**

FIGURE 3-6: **Start a new piece of yarn the same way you started the first piece.**

FIGURE 3-7: **Wrap the new yarn over the ends of both pieces of yarn.**

FIGURE 3-8: **To build the sides, lay the next row right on top of the last one.**

FIGURE 3-9: **Keep adding rows to build the sides up.**

9. When the sides are high enough, finish off the basket by securing the end of the core. Cut the end of the core at an angle so it tapers down. Wind it tightly with wrapping and take a few stitches to hold it securely. With the needle, poke the end of the wrapping back into the basket so it doesn't show.

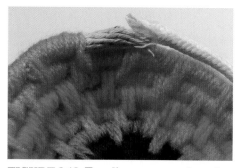

FIGURE 3-10: **To splice on a new piece of core, match the angle of the cut to the old piece.**

FIGURE 3-11: **Cut the core at an angle to end it.**

FIGURE 3-12: **Pull the yarn through some stitches to hide the end.**

Extension:

To make handles on the top coil, pull a section of the top coil away from the basket to make a loop. Wrap the loop separately. Then connect the other end to the basket and continue the top coil. Repeat on the other side of the basket.

FIGURE 3-13: **Handles on a basket**

WEAVING ON LOOMS

Weaving is a way to make textiles from two interlocking sets of thread, one vertical and one horizontal. Archaeologists have found carved stone figures wearing woven skirts and belts dating back more than 20,000 years. Modern methods are not that different from those used in ancient times. Weaving is done on a *loom*, which is a frame or machine that holds the threads in place and sometimes helps move them in and out. *Warp* threads usually go up and down. Although they may not even be visible on the finished fabric, they give the woven cloth its strength and stiffness. The *weft* (also called the *woof*) threads go in and out between the warp threads. The color and texture of the weft threads, and the way they are woven in and out of the warp threads, are what give the fabric its look and feel.

For thousands of years, weaving was done by hand, usually in the home by women and girls. But as you learned in Chapter 1, in the 1800s the Industrial Revolution brought with it automated looms and textile factories. These turned fabric from an expensive luxury to a common item even ordinary people could afford. Today weaving by hand is done to express creativity, and handmade fabrics are an art form instead of a necessity.

FIGURE 3-14: **A weaving on the loom where the ends have not been woven in**

✂ FABRIC AND FIBER INVENTORS: JOSEPH MARIE JACQUARD

In 1801, a French weaver named Joseph Marie Jacquard introduced a new invention: a device that could create intricate designs on woven fabric automatically. The device controlled which threads were used on the loom for each row—a job that up until that point had been done by a person working with the weaver. The instructions for Jacquard's loom controller were printed in the form of cards with holes punched in them. Each combination of holes resulted in a different combination of threads, and multiple cards could be linked into a chain to produce an entire picture. To demonstrate his new loom, Jacquard used it to produce a life-size portrait of himself.

Jacquard's idea changed the textile industry, but it also influenced other areas of exploration. In 1837, a British mathematician named Charles Babbage came up with the idea for a primitive computer that could take information and process it. To load the information into the machine, Babbage borrowed Jacquard's idea of punch cards with holes in them. Babbage never actually built his "analytical engine," but his idea inspired a young mathematician named Ada Lovelace to design a way to tell Babbage's machine what to do with the information it was given. Her idea led directly to the invention of computer programming—and it was all inspired by a weaving machine.

PROJECT: CARDBOARD WEAVING LOOM

⦿ MATERIALS

Stiff sheet of cardboard, at least 1 inch (2.5 cm) bigger on all sides than the finished product (a 6-inch [15-cm] square makes a coaster-sized weaving)

2 strips of cardboard about ½ inch (1.25 cm) wider and the same length as the width of the loom

White glue

Yarn, as many colors as you like

Tape

(Optional) sticks or rods for hanging your weaving

⦿ TOOLS

Ruler

Pencil

Yarn needle

(Optional) metal fork or wide-toothed comb

(Optional) small- to medium-width crochet hook

FIGURE 3-15: A coaster made on a cardboard loom

To make a quick, temporary loom, just grab some scrap cardboard. You can practice the basics of weaving by creating a set of coasters, a placemat, or a wall hanging.

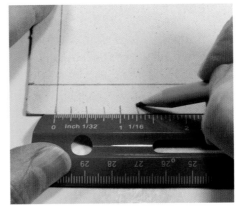

1. First, make the loom. To do so, draw a line ½ inch (1.25 cm) away from each edge of the cardboard. Along the top and bottom, make marks every ¼ inch (75 mm). Cut slits at each mark. These are the notches to hold the warp thread on the loom.

FIGURE 3-16: **Marking the notches**

2. On the top and bottom of the loom, glue the strips of cardboard along the lines, away from the edges. These will lift the warp threads and make them easier to work with.

FIGURE 3-17: **Cutting the notches**

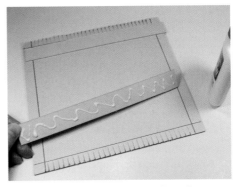

FIGURE 3-18: **Gluing the cardboard**

3. You *dress* the loom by wrapping the warp threads around it. Here are the steps:

A. To start, tie a knot in the warp yarn, leaving some extra as a tail. Slide the yarn into the first notch on the bottom, from back to front (see Figure 3-19). Tape the tail to the back of the loom.

FIGURE 3-19: **Slide the yarn through the first notch from front to back.**

B. Next, take the yarn and slide it into the first notch at the top of the loom. The yarn should be tight but not bend the cardboard (see Figure 3-20).

C. Bring the yarn around the back of the loom and slide it through the next notch on the bottom.

D. Then bring it up the front to the next notch on the top.

FIGURE 3-20: **Thread the yarn through the first set of notches.**

Continue wrapping the yarn around the loom the same way through all the notches, ending with the yarn at the bottom of the loom. Cut the yarn, leaving another tail, and tape it to the back of the loom (see Figure 3-21).

4. Start the first row just inside the cardboard spacer (either end you choose). To begin weaving, take about 3 feet (1 m) of the

FIGURE 3-21: **The back of the loom after it is dressed. The tails at the beginning and end are taped down to secure them.**

weft yarn and thread it on the needle. Bring the needle under the first warp yarn, then over the next (see Figure 3-23). Continue weaving the weft yarn in and out through the warp threads until you reach the end of the row. Pull the yarn through until there is just a small tail left hanging off the side (see Figure 3-24).

FIGURE 3-22: **The front of the loom when it is dressed**

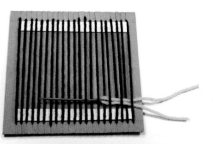

FIGURE 3-23: **Bring the needle and the yarn in and out of the warp threads.**

FIGURE 3-24: **The first row**

5. Start going back the other way, but make the stitches the opposite of the way you wove them in the first row. In other words, if you end with the yarn going under the last warp thread, the first stitch of the next row goes over that same warp thread. This time, curve the yarn instead of going straight across.

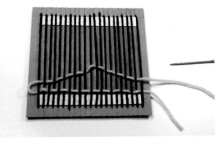

FIGURE 3-25: **Leave a hump in the next row so there's a little extra yarn.**

The "hump" you create will give you a little extra yarn to keep your rows from getting too tight and pulling in on the sides. Continue making rows with humps in them.

6. Every couple of rows, push the rows together so the warp threads don't show as much. This is called *beating*. You can use your fingers or a fork or comb. The tighter the rows, the smoother your weaving will look.

FIGURE 3-26: **Beating the yarn with a fork**

FIGURE 3-27: **Second row after it has been beaten flat**

7. Repeat Steps 5 and 6 until your weaving is as long as you want it to be. Stop before you get to the cardboard spacer strip.

8. To finish your piece, turn the loom over. Cut the warp threads right across the middle.

9. On the front of the loom, pull the first two warp threads on the bottom loose. Tie them together once.

FIGURE 3-28: **Cut the warp threads right down the middle.**

10. Then make an overhand knot by taking both ends, making a loop with them, and then pulling their ends through the loop. This will help the fringe lie straight. Repeat this process with the rest of the warp threads. If you have a warp

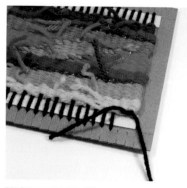

FIGURE 3-29: **Tie two loose ends together.**

thread left over at the end, tie it around the warp yarn. Trim the ends as needed.

FIGURE 3-30: **Curl both ends into a loop and pull them through it.**

FIGURE 3-31: **The finished knot looks very neat.**

11. Now it's time to cut off any leftover tails of weft yarn sticking out that are already anchored. If they have not already been anchored, thread them on the needle and weave them into the back or sides of the weaving. Make sure they go through enough stitches to keep them in place, then trim their ends so nothing sticks out.

FIGURE 3-32: **Work loose ends into the back of the weaving (the side facing you as you worked).**

FIGURE 3-33: **Clip the ends so they don't show.**

12. Your weaving is finished! Now that you have the hang of it, make more using some of the following suggestions. If you are careful, you should be able to reuse your cardboard loom for several more weavings.

FIGURE 3-34: **A finished weaving**

✂ WEAVING TRICKS AND SHORTCUTS

Instead of leaving all the tails of weft yarn hanging off the sides (as shown in Figure 3-14), you can weave them in as you go and save "clean-up" time later.

To start the piece, follow these steps:

1. Before you weave the first row, anchor the yarn by weaving in a few stitches going from the middle to the edge. The stitches will sit in between where you plan to make your first and second rows. Leave a short tail hanging off in the middle, as shown in Figure 3-35.

2. Below the anchor stitches, weave the yarn in and out across the entire row, as described earlier in Step 4 of the main text. (See Figure 3-36).

3. To make the second row, follow the directions earlier in Step 5 of the main text, this time going above those anchor stitches. When you beat down the rows, the anchor stitches will be locked between the first two rows. Try to push the rows above and below the anchor stitches together tightly so they stay straight. (See Figures 3-38 and 3-39 on the following page.)

FIGURE 3-35: Start by weaving a few stitches between the first row and where the next row will go. Leave a short tail hanging off the front of the piece.

FIGURE 3-36: Weave the first row below the short tail.

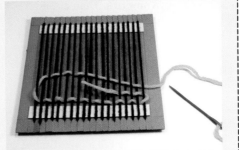

FIGURE 3-37: Weave the second row above the tail.

FIGURE 3-38: Lock the tail between the first and second rows.

FIGURE 3-39: The first two rows beaten down

To change yarn in the middle of a row: Cut the old yarn, leaving a short tail. Take the needle with the new yarn and weave a few stitches that overlap the short row—following the same in-out pattern as the last full row. Finish the row with the new yarn. Pull the new yarn almost all the way through until there is just a short tail left and let the tail hang off the front (see Figure 3-40). Continue weaving as before.

FIGURE 3-40: To change yarn in the middle of a row, overlap a few stitches at the ends.

To change weft yarns at the end of a row: Weave a few stitches with the old yarn as if you are starting a new row. Then cut it, leaving a short tail. On the opposite side of the weaving, weave a few anchor stitches going from the middle toward that side of the row. Pull the new yarn through until you have just a short tail sticking out. (See Figure 3-41.) You should have one tail a few stitches in from each side.

FIGURE 3-41: To change yarn at the end of a row, weave a few anchor stitches going from the middle toward the end. Then weave the next row as usual.

Now, weave your first full row with the new yarn above the two tails so they are locked between the row of old yarn and the row of new yarn. Continue the next row as usual.

Bonus tricks:

- ☑ If you are changing colors and repeating them every two or three rows, don't cut the first yarn when you finish. Just leave the entire end hanging with the needle still attached and continue with a new needle and your second color. When you're ready to use the first color again, take the needle with the first yarn attached and insert it through the stitches directly above it until you reach the row where you want to start using it again.

FIGURE 3-42: **This example uses the same piece of silver yarn in every other row.**

- ☑ To make an "island" of yarn surrounded by other yarn, first weave the island section on the bare warp threads, leaving hanging tails at both ends. Then fill in around your island with the other yarn. Where the second yarn touches the first, hook the two sections together. Do this by bringing the second yarn through the stitch at the end of the yarn island.

FIGURE 3-43: **If you have islands of color, make them first, then link them to the yarn around them.**

Adaptations and extensions:

- ▣ For younger weavers, cut the slits in the loom farther apart.

- ▣ To create loops that you can use to hang your weaving, use a more traditional way of dressing the loom:

    - ⊙ Slide the warp yarn into the first notch on the bottom from back to front.

    - ⊙ Then bring the yarn up to the first notch at the top of the loom and slide it through from front to back.

    - ⊙ Bring the yarn across the back of the tab to the next notch over and slide it through the notch from back to front.

    - ⊙ Carry the yarn down to the next open notch on the bottom and repeat, until the loom is fully dressed.

- ▣ To create narrow vertical stripes with two different weft yarns, weave one row with the first yarn and the next with the second yarn, then repeat. Only the "over" stitches will show, creating columns.

- ▣ Weave in thicker yarn, t-shirt yarn, or even strips of fabric to create a piece with bumps and interesting textures.

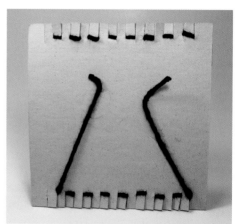

FIGURE 3-44: The back of a loom where the warp threads can be used for hanging the piece when finished

FIGURE 3-45: Loops are created when you remove the weaving from the loom.

FIGURE 3-46: Pull the top loops to close up the bottom loops.

BUILDING WITH FABRIC

Fashion and craft designers aren't the only people who use fabric and fiber. Architects and engineers turn to soft materials like fabric when they want to create unusual shapes or cover large areas like ceilings without adding a lot of weight. No matter what the material, the process is the same: start with a problem to solve, come up with ideas, do research to come up with the best option, create models and prototypes to test your idea out—then go back to the drawing board to make the design even better and keep testing it until you've got the best solution possible.

Sometimes the problem can be solved with advanced textiles. Sometimes the answer involves using common materials in unusual ways. In the following project, you'll use a special camping goods fabric called ripstop nylon to create a wearable shelter. As you work on it, think about ways you could design a better version of your own!

✂ FABRIC AND FIBER INVENTORS: VERONIKA SCOTT AND ANGELA LUNA

Two design students have invented new kinds of transformable coats to solve problems they saw in their community and around the world—they've gone on to start their own companies to make their ideas a reality.

Veronika Scott (www.empowermentplan.org/) was a student at the College for Creative Studies in Detroit when she was given an assignment: create a product to fill a need in the community. She came up with the idea for the EMPWR Coat, a warm, waterproof coat that becomes a sleeping bag and can be carried like a shoulder bag when not in use. But when Scott interviewed homeless people at shelters in her area to get their feedback on her creation, she found that they wanted jobs as much as coats. So in 2012, she started a nonprofit company called The Empowerment Plan. Scott's company hires homeless parents and trains them to work running sewing machines. The EMPWR Coat is made with fabric donated by Carhartt, a company that makes work clothes, and upcycled insulation from car-maker General Motors. So far, The Empowerment Plan has given away over 15,000 coats to people in need throughout North America.

Angela Luna (www.adiff.com/) studied fashion design at Parsons School of Design in New York, but in her senior year, she decided to do more than just make clothing that looked nice. Hearing about the problems faced by refugees traveling to Europe, she asked herself, "Where is there a design problem that we could offer a solution to?" Her solution was a series of coats that could turn into tents, sleeping bags, flotation devices, and more. After graduating from Parsons in 2016, where she won the Womenswear Designer of the Year award, she formed a company called ADIFF—its mission is to "make a difference." Luna took her prototype coats to Greece and asked refugee families to test them out. Then in 2017, ADIFF launched a Kickstarter campaign for a reflective coat that can be seen in the dark. They raised almost $100,000. The company plans to use some of that money, plus profits from selling ADIFF coats to people who like the way they look, to donate coats to refugees in need.

PROJECT:

WEARABLE SHELTER

⊙ MATERIALS

4 yards (4 m) ripstop nylon, about 60 inches (1.5 m) wide

Wooden pole (about 5 feet [1.5 m] high)

9 tent stakes

(Optional) Velcro fasteners

⊙ TOOLS

Piece of string a little longer than 5 feet (1.5 m)

(Optional) Iron (make sure to iron on a very low setting, or use a towel or cloth under the iron) and ironing board

Pencil or fabric chalk

Masking tape

Pins

Sewing machine

NOTE: All the sewing for this project is done on a basic sewing machine. You should know how to sew simple seams before starting. Children should get help from an adult. See "Using a Sewing Machine" in the Introduction.

FIGURE 3-47: A wearable tent prototype

Take on the challenge of creating a wearable tent! Here is one possible solution—a rain cape that can be set up as a tipi-style shelter.

1. Cut the fabric to a length of 10 feet (3 m). Set the rest aside to use later for the hood. Unfold the fabric and lay it out flat. (Press it with a cool iron if needed.) Fold the piece of fabric in half so the shorter ends meet.

FIGURE 3-48: **Fold the fabric in half.**

2. Choose one corner of the fold to be the top of the cape/tent when it is finished. Then you need to fold the fabric a few times to form a skinny triangle. To start, take the corner of the fold marked Bottom Corner #1 in Figure 3-49 and bring it up to meet its opposite corner. It will begin to look like a triangle.

3. Fold the fabric again, but this time take Bottom Corner #1, along with the layers of fabric underneath it, and fold them down so they line up with the diagonal (tilted) fold you created in Step 2. You now have another corner, Bottom Corner #2.

FIGURE 3-49: **Fold the fabric into a triangle.**

4. Stretch a string along the folded edge of the fabric between the top corner and Bottom Corner #2, leaving extra string at each end. Tape one end of the string to the top corner, as shown in Figure 3-50. Tie a pencil to the

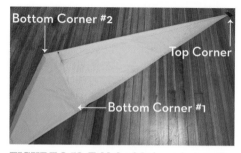

FIGURE 3-50: **Fold the fabric again to make a thinner triangle.**

other end of the string. Keeping the string tight to guide you, take the pencil and draw a curve from that corner across to the other folded corner. Cut along the curved pencil line. Save the leftover fabric to make tabs.

5. Unfold the fabric into a semicircle. To make a hem along the straight edge of the fabric, fold over 1 inch (2.5 cm) of the edge, then fold it over again. Press the fabric flat with a cool iron if needed. Pin and sew the hem.

6. Take the hood fabric you set aside and fold it in half so it is roughly square, right sides together (if your fabric has two different sides). Trim the folded fabric to about 15 inches (38 cm) on each side. (Make the hood bigger or smaller to fit your head.) With the fold at the top, pin one side closed and sew it with a ⅝ inch (1.5 cm) seam allowance. This is the back of the hood.

7. Clip the top corner of the seam allowance at the fold on an angle. Press the seam open and turn the hood right side out. Sew a narrow hem along the bottom of the hood.

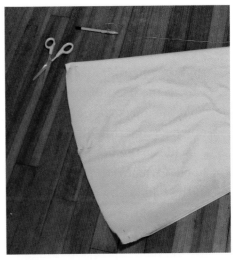

FIGURE 3-51: **Use the pencil and string to draw a curved line, then cut along the line.**

FIGURE 3-52: **When you unfold the fabric, you will have a semicircle.**

FIGURE 3-53: **Sew the back of the hood.**

8. Lay the hood out flat. Fold the cape in half so the corners meet. Lay the cape out flat so that the opening faces the same way as the front of the hood. Fold the top corner of the cape down so that the fold is almost as wide as the bottom of the hood. Measure the short sides of the folded down fabric to check that they are equal. (In Figure 3-54, the short sides are 8.5 inches [21.5 cm] long.) Cut along the fold to make the neckline.

FIGURE 3-54: **Fold down the top of the cape to match the bottom of the hood.**

9. Fold down ½ inch (1.25 cm) along the cut to make a narrow hem. Sew along the middle of the hem. Pin the bottom of the hood to the neckline, right sides together. Match the front edge of the hood to the front opening of the cape. Match the back seam to the back fold of the cape. Sew. (See Figure 3-57.)

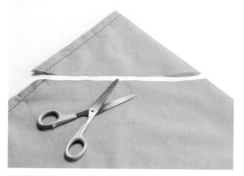

FIGURE 3-55: **Cut along the fold.**

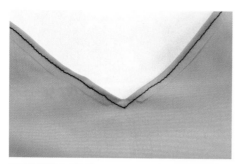

FIGURE 3-56: **Sew a narrow hem along the neckline.**

FIGURE 3-57: **Attach the hood to the neckline.**

10. Take the fabric you set aside to make tabs. Cut nine tabs that are 2 inches by 6 inches (5 cm by 15 cm). To make each tab, fold the fabric in half so the long sides meet, and press. Unfold, then fold each half so the raw edges meet in the middle. Fold along the center fold with the raw edges inside. Press again. Sew the tab up the middle.

11. Fold over 1 inch (2.5 cm) of the curved edge of the cape, then fold it over again. Press. Pin the hem closed. Take the tabs and fold them in half so the raw edges meet. Insert the tabs, raw-edge side down, inside the hem at the front edge, in the fold at the back, and evenly between those points. Pin the tabs in and sew the hem closed. When you reach a tab, sew over it, sew backward over it, and then sew forward over it again.

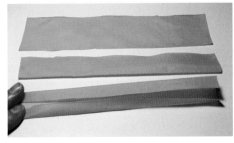

FIGURE 3-58: **Fold each tab in half, twice, so the raw edges meet inside.**

FIGURE 3-59: **Sew each tab closed.**

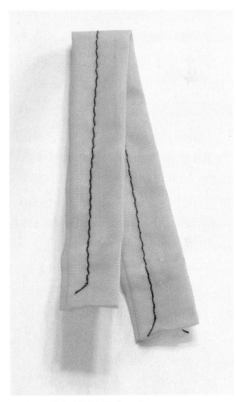

FIGURE 3-60: **Fold the tab.**

12. Your wearable tent prototype is
ready for testing! To wear it as
a rain cape, attach round Velcro
fasteners or sew on thin strips
that you can tie together at the
neck opening. To set it up outside as a tent, stick a pole in the
ground. Drape the hood over
the top of the pole, and spread
the bottom of the cape out as
far as possible. Put a tent stake
through each tab to hold the tent
in place. As you try it out, think
about ways to improve it. See the
following extensions for some
ideas.

FIGURE 3-61: **Insert the tab into the hem.**

Extensions:

☐ Cover the front of the hood with
detachable mosquito netting to
let in light and air.

☐ Add inside pockets to hold your
belongings when you're sleeping
in the tent.

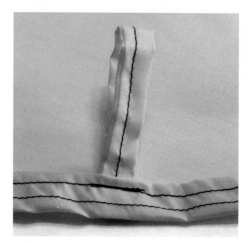

FIGURE 3-62: **Sew backward and forward over each tab to make sure it is secure.**

☐ Sew on a pouch with straps so
you can fold up your wearable tent and carry it like a backpack.

☐ Try out some ideas to make the tent bigger and still wear it as a cape,
such as adding a flap in the back that folds up when you're wearing it.

☐ Look at traditional tents for ways to make the cape/tipi more waterproof, like covering the seams with tent sealer.

☐ Come up with your own design for a coat that transforms into something useful. You can use paper and cardboard to make a doll-sized
model to test out your idea, then look around for material you can use
to build a full-sized prototype, such as an old shower curtain, a plastic
tarp, or a vinyl tablecloth.

THE MATH OF QUILTING

It's hard to make a quilt without geometry. A quilt is a blanket or other fabric decoration with three layers: a top made of patches of fabric sewn together, a middle of fiber stuffing, known as *batting*, and a bottom piece of fabric that is usually solid. To make the top of a quilt, you have to *piece* different shapes of fabric together, a little like doing a jigsaw puzzle. However, the fabric pieces are usually geometric shapes like squares, rectangles, and triangles. By using just a few simple shapes in different colors and designs, you can create thousands of different patterns.

To hold the quilt together, and to keep the stuffing from moving around, you stitch right through all the layers. Sometimes the stitching, which is known as the *quilting*, simply follows straight lines. But you can use the quilting to create a separate pattern that can add even more complexity to the quilt equation.

Quilters are clever people who like to figure out shortcuts to make interesting patterns without having to sew all the patches together one at a time. The checkerboard quilt in this chapter uses a technique called strip piecing, so it's much easier to make than it looks.

NOTE: Part of the fun of making a quilt is searching for fabric in interesting patterns and colors that go together well, but you can also try using some of the patterned and painted fabric you created in other projects in this book. It's a great way to recycle any "failures" that didn't come out the way you wanted!

PROJECT: A FOLDABLE QUILTED CHECKERBOARD

◉ MATERIALS

4 strips cotton quilting fabric, 2½ by 42 inches (roughly 6.5 cm by 107 cm), two strips in one color or pattern and two in a different color or pattern so they are easy to tell apart (strips of fabric in that size are often sold in rolls, or you can cut 2 strips each out of ½ yard [50 cm] of bolt fabric)

Pins (straight or safety pins)

1 piece thin cotton batting, 16½ inches (42 cm) square (the same size as the finished top of the quilt)

1 piece of cotton fabric for the bottom, at least 18 inches (46 cm) square (can be cut from 1 "fat quarter," (a precut piece of quilting fabric that is about 18 by 22 inches [46 by 56 cm], or from ½ yard [50 cm] of bolt fabric—for a wider edging, use a bigger square)

Masking tape

◉ TOOLS

Sewing machine

Iron and ironing board

Seam ripper

Hand sewing needle (thin or quilting needle)

Medium safety pins

1 or 2 thimbles

Ruler

Chalk

FIGURE 3-63: A quilted chess board with game pieces

NOTE: For the meaning of specialized sewing terms, see the Introduction.

This soft checkerboard lets you take your game wherever you go. It folds up easily along the quilted seams. You need to create an 8×8 grid of squares with different-colored alternating squares—but to make all those squares, you only need to sew a few long strips. Magic! The squares on the checkerboard will be 2 inches (5 cm) on each side, big enough to fit most checker or chess pieces.

1. First, you need to sew the four strips of fabric together, alternating the two colors. To start, take one strip of each fabric and lay one on top of the other, right sides together. Pin along one long side and sew the strips together with a ¼-inch (6-mm) seam allowance. Repeat with the other two strips.

2. Next, sew the two pairs of strips together the same way, with opposite colors touching. Press the seams toward the darker fabric. This is your strip set.

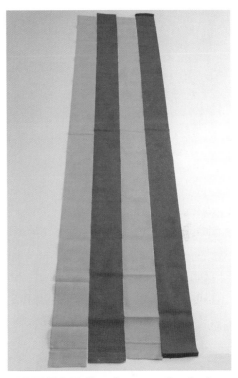

FIGURE 3-64: **Start by connecting the four strips.**

FIGURE 3-65: **Sew the strips together in pairs.**

FIGURE 3-66: **Sew the two pairs of strips together.**

FIGURE 3-67: **The four strips sewn together**

3. Fold the strip set in half so the shorter edges are lined up. Cut in half along the fold. Sew the two new pieces together so opposite colors are touching. Press the seam toward the darker fabric.

4. Use a ruler and chalk to mark eight equal rows, each 2½ inches (about 6.5 cm) wide, down the length of the fabric (see Figure 3-70). Cut across the fabric to create eight columns.

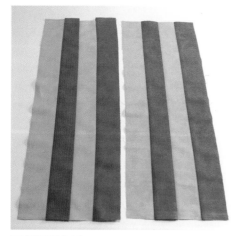

FIGURE 3-68: **Fold and cut the long strips in half.**

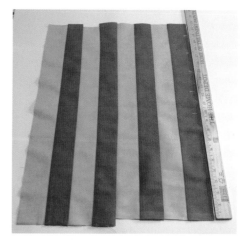

FIGURE 3-69: **Sew the two halves together.**

FIGURE 3-70: **Cut into rows.**

FIGURE 3-71: **Flip every other row so the colors alternate.**

5. Flip every other row so that the opposite squares of color line up. Pin, trying to match seams, and sew.

FIGURE 3-72: **How the rows should line up**

6. Before you go further, check the back to make sure all the seams are sewn flat. If you see a twisted seam, use a seam ripper to open enough stitches to straighten it out. If it leaves a large hole, close it up with a few hand stitches. Small holes will be hidden by the quilting.

FIGURE 3-73: **Top sewn**

7. Next you will stack the layers of your quilt. Press your backing fabric with the iron if needed. Place the backing—right side down—on a table or board. Tape each corner down, pulling the fabric to make it lie flat (but do not stretch it). Center the batting on the quilt bottom. Smooth it out with your hands. Then center the quilt top on the batting.

FIGURE 3-74: **Use a seam ripper to undo any mistakes.**

FIGURE 3-75: **The back of the quilt top, with all seams fixed and pressed flat.**

8. Now you will *baste* the layers using safety pins to hold them together while you work. Start in the center of the quilt and work your way out to the edges. Use plenty of pins! Avoid pinning over the seams so that you can leave the pins in as you stitch the quilting. When you're done, untape the quilt bottom and look everything over to make sure it is as smooth as possible.

FIGURE 3-76: **Tape the quilt bottom down.**

FIGURE 3-77: **Place the batting on top of the bottom.**

9. Time to quilt your layers! The style used on your checkerboard is called "quilting in the ditch" because you sew right over the seams on the top piece. Each row is hand-stitched separately, beginning with the middle seams in both directions. As you do the remaining rows, keep smoothing the fabric toward the edges to avoid wrinkles. To start each row, thread a needle and tie a small knot in the longer end of the thread. Poke the needle through the seam allowance and pull it gently until the knot stops it from going through. Then poke the needle up right through the seam, near the edge.

FIGURE 3-78: **Pin all the layers together.**

FIGURE 3-79: **Take a stitch in the seam allowance only to secure the knot.**

FIGURE 3-80: **Poke the needle from the back to the front of the top, right through the seam.**

10. Now use a running stitch to go along the seam. Make sure to catch all three layers! When you get to the end, anchor the thread by sewing through the seam allowance and tying another knot. Cut the thread. Do the same with the middle seam going in the other direction. These two lines of stitching that cross in the center of the quilt will help hold the fabric sandwich in place. Repeat with the remaining seams, working from the middle toward the edge.

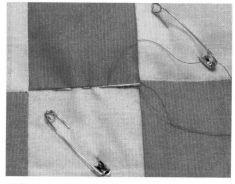

FIGURE 3-81: **Sew along the seam with a medium running stitch.**

FIGURE 3-82: **The first two rows of quilting down the middle seam in both directions**

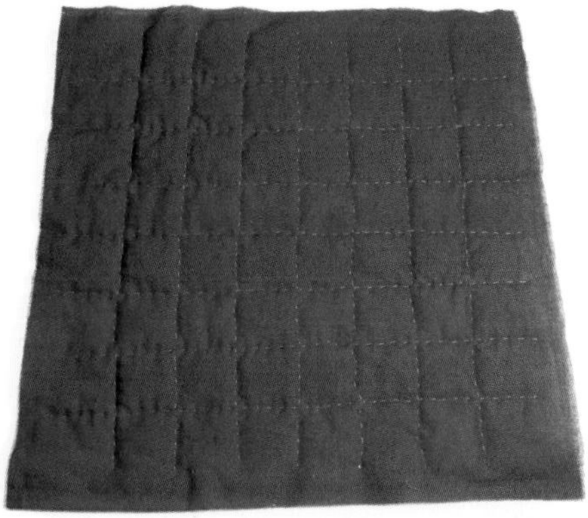

FIGURE 3-83: **The finished quilting as seen from the bottom of the quilt**

11. When you are done with the quilting, you will finish the edges by folding the extra bottom fabric over twice. The edging will overlap the squares along the side by ¼ inch (6 mm). That means the extra fabric from the bottom should wrap around your top by about ½ inch (1.25 cm) all the way around. If it's very uneven, trim it.

FIGURE 3-84: **Close-up of the finished quilting**

FIGURE 3-85: **Trim any batting sticking out.**

FIGURE 3-86: **The quilt ready for hemming**

12. To start forming the edging, take one side and fold this extra in half so the edge of the bottom meets the edge of the batting and top piece. Press it flat with the iron. Then fold it over again along the top edge, so that the bottom is now overlapping the top. Press again and pin to hold in place.

FIGURE 3-87: **Fold one hem in.**

13. At the bottom corner, fold the fabric at an angle along the next side (see Figure 3-88). Then fold the next side down like you did in Step 12. Continue until all four sides are pinned. Once the pins are in place, use a running stitch down the middle of the edging to hold it in place.

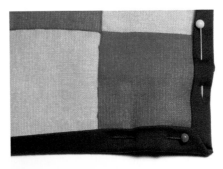

FIGURE 3-88: **Fold the corner as shown.**

FIGURE 3-89: **Fold the second hem up, matching the angle of the corner.**

14. Your board is finished! Find some checker or chess pieces and test it out. Then try some of the following extensions.

FIGURE 3-90: **The finished board**

FIGURE 3-91: **The top and bottom of the finished board**

Extensions:

- ▣ Add Velcro straps to secure the board when it is rolled up.
- ▣ Make a stuff sack to hold the game pieces.
- ▣ Make stuffed fabric chess or checker pieces. For travel, use a fluffy fabric for the squares, and stick or sew on Velcro to the bottom of the pieces.
- ▣ Design a board quilt for your favorite game—or make up your own game.

✂ FABRIC AND FIBER INVENTORS: HARRIET RIDDELL

British artist Harriet Riddell (instichyou.co.uk) makes drawings with thread instead of a pencil. She first started stitching scenes and portraits with her sewing machine at the suggestion of an art teacher. The "one-line" style fits with her style of drawing, she believes. "You just flow with the needle." And because she hates being indoors, she has figured out a way to take her sewing machine on the road—by bicycle!

When Riddell began carrying her sewing machine around London, she had trouble finding places to plug it in. Then an arts festival supplied her with a bicycle that had an electric generator. Today, Riddell totes her machine around in a trailer attached to her own bicycle. Her subjects pedal while they pose. A display tells them how much energy they are generating.

The set-up lets her models interact with the work while it is being created. When she's sewing something complicated with lots of tight stitching, they have to pedal harder as the motor does more work. The artist also asks her subjects to tell her their stories. She often adds their words to the final design. "I like that people feel a part of the art," she says.

Riddell's portable sewing machine has traveled as far as Kenya and India. She has sewn while being driven down the street during a festival and stitched the sunrise in the Himalaya mountains. But mostly she sticks closer to home, drawing the city landmarks and interesting characters she comes across using people power and imagination.

FIGURE 3-92: Harriet Riddell stitching a scene in front of a crowd in India CREDIT: DAVID TRATTLES

NEEDLE ARTS

These easy projects will help you get started with knitting and crochet, but be careful— they can be habit-forming!

Just like other kinds of fabric and fiber art, knitting and crochet rely heavily on math and engineering. And they're about more than just counting and measuring—needle arts can be used to demonstrate all kinds of interesting math concepts. For instance, sewing a tent involves taking a flat, two-dimensional piece of material and turning it into a three-dimensional structure with height and width and depth. With knitting and crochet, it's almost like taking a one-dimensional line—a strand of yarn—and making it into a three-dimensional piece of clothing. In fact, some people even use knitting and crochet (and other kinds of needle arts, like embroidery and cross-stitching) to help explain how computers work. In computer programming, you input data, process it using a computer program, and end up with a result called the output. In knitting, you input the yarn, process it on your knitting needles following a pattern, and end up with a result called a sweater (or mittens, or a scarf).

Knitting is also *binary* like a computer, meaning there are only two conditions. Inside a computer, every piece of information is reduced to either 1 or 0, on or off. With basic knitting, every stitch is either a *knit* or its opposite, called a *purl*. (You'll learn what those terms mean later in this chapter.)

A computer program is a special language, or code, with its own terms and way of saying things. Needle arts also have their own kind of code. Instead of writing out every instruction in sentences, patterns for experienced knitters are written using abbreviations. You can even save time with both computer code and knitting code by using repeated blocks of instructions to do the same series of steps over and over.

In this chapter, you'll learn some of the code used in knitting and crochet patterns and the techniques for processing the input (also known as stitches). Once you understand it, you'll be able to figure out how to make other designs using the same "language"—or even make some of your own!

✂ NEEDLE ARTS TIPS

- ▣ A *skein* of yarn looks like a tube. It is usually wound so that the ends are tucked into the hole in the center of the tube. If you loosen both ends, you should be able to pull the yarn you're working with from the center, which keeps the skein from rolling around as you work.

- ▣ A ball of yarn is wound so that one end is buried in the middle and the other is on the outside. To keep a ball of yarn from rolling around while you work, place it in a basket on the floor, table, or chair next to you. To carry your work with you, place it in a zip-top bag. When you want to use the yarn, open the bag just enough to let the strand of yarn slide through.

- ▣ Before you start to use a new technique in a project, practice it by making a *swatch*, or small sample piece. It can be just five or six rows with about 10 to 20 stitches per row.

- ▣ Counting stitches is important for some patterns. However, for the projects in this book, you can adjust the width of your piece as you go along without worrying about counting. If you accidentally drop or pick up an extra stitch, just add or subtract stitches as needed in the next row. (See the instructions later in this chapter for how to increase or decrease.)

- ▣ To undo a few stitches or an entire row, take out the crochet hook or knitting needle and pull on the yarn—but go slowly so you don't pull out more rows than you mean to.

- ▣ In some patterns, you will see a listing for the *gauge*, which tells you the number of stitches per inch (or cm) the project should have. The number will vary depending on how tight or loose you make the stitches. For the projects in this chapter, you can adjust the thickness of the yarn, the size of the knitting needles or crochet hooks you use, or even the number of stitches you make to get the look you want without worrying about a specific gauge.

- ▣ If you are left-handed, you may want to switch any directions for "left" and "right." However, knitting and crochet use both hands nearly equally, so you may be fine following the standard directions.

- ▣ For places to find specialized materials, see "The Fabric and Fiber Supply Basket" section at the beginning of this book.

If knitters are the math nerds of the craft world, says audio expert and kinetic artist Jesse Seay (jesseseay.com), then machine knitters are the engineers. Seay is a professor at Columbia College in Chicago where she teaches audio production and sound art. She also creates yarn-based electronics and light-up designs using vintage knitting machines.

Seay doesn't consider herself a knitter. "I got interested in knitting machines because I like machines," she says. "They are very complex." For her, the knitting machine is a tool that can be used in a new way to support her passion for technology.

One thing Seay discovered about knitting is that the design is built into the structure. For example, take Fair Isle sweaters—the kind that have knitted-in patterns of snowflakes or other wintery images. The Fair Isle style results from a special knitting technique that lets you use different colored yarns in the same row. And knitting machines are very good at making Fair Isle patterns.

However, finding machines and learning to use them took some research. Hand-powered home knitting machines were popular in the

FIGURE 4-1: Jesse Seay working on her knitting machine CREDIT: SABRINA RAAF

FIGURE 4-2: Fair Isle sweater folded up to show the connecting stitches on the wrong side of the fabric

1970s and '80s, but they were complicated to use and control. Some models used the same kind of punch cards used by programmable textile machines almost 200 years earlier. By the end of the 1990s, interest almost disappeared, and machines became harder to find.

Today, Makers and hobbyists are helping to revive home machine knitting. And they're turning to local machine knitting clubs and online machine knitting interest groups on sites like Ravelry (www.ravelry.com) for information. Seay drew on sources like these to figure out how to combine wire and cotton yarn in a Fair Isle pattern to knit her own electronic circuit boards and create other designs, such as knitted stretch sensors you can wear like the finger of a glove.

Now Seay shares her techniques and discoveries in her classes and workshops, on her blog, and on websites like Instructables (www.instructables.com). "My real goal," she explains, "is getting knitters into electronics." She's hoping the hardware enjoyed by past generations of women crafters will inspire the next generation of women engineers.

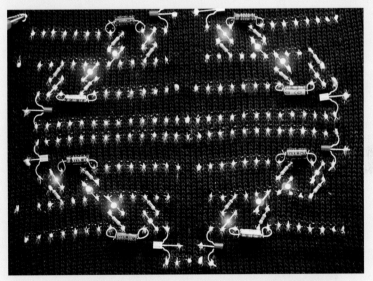

FIGURE 4-3: A Fair Isle electronic circuit made on a knitting machine. It uses wire instead of yarn for its second "color." Seay has added a dab of solder—a kind of metallic "glue"—to each wire stitch. More solder is used to attach components like LED lights.

CREDIT: JESSE SEAY

QUICK CROCHET HOW-TO

These are the basic techniques you'll need to know to make the crochet projects in this chapter.

Slip Knot

The slip knot is used to start a new project. Make a loop in the yarn, about 10 inches (25 cm) from the end, with the working end of the yarn (the end closest to the skein) on top. Bring the working end up through the bottom of the loop, making another loop. Pull the first loop tight. Insert the hook into the second loop and pull one end of the yarn until the loop is secure on the hook but not too tight.

FIGURE 4-4: **Make a loop in the yarn, with the working end of the yarn on top.**

FIGURE 4-5: **Bring the working end up through the bottom of the loop, making another loop.**

FIGURE 4-6: **Pull the first loop tight.**

FIGURE 4-7: **Insert the hook into the second loop and pull it until it is secure but not too tight.**

How to Hold the Yarn and Hook

Hold the yarn and the hook in whichever hand is comfortable for you. To control the *tension* of the yarn (how tight it is), wrap the working end of the yarn (closest to the skein) around the pinky and first finger of the hand without the hook. As you crochet, hold the piece between your thumb and second finger on that hand.

FIGURE 4-8: **Wrap the working end of the yarn (closest to the skein) around the pinky and first finger of the hand without the hook.**

Yarn over

Bring the hook behind the yarn and let the yarn slide into the working part of the hook.

Chain (ch)

A chain is the foundation for every new crochet project. Make a slip knot, then yarn over with the working end of the yarn, and pull the yarn through the loop on the hook. Con-

FIGURE 4-9: **Performing a yarn over**

tinue making loops as many times as you need to. Each V-shape in a chain counts as one chain stitch.

FIGURE 4-10: **Pull the yarn through the loop on the hook.**

FIGURE 4-11: **Continue making loops as many times as needed.**

Single crochet (sc)

This is the basic crochet stitch. To begin, first make a chain. Skip the first V-shaped chain stitch next to the hook. Stick the hook into the middle of the second chain stitch. Yarn over and pull the loop of yarn back through to the top. You now have two loops on the hook. Yarn over again, and pull the new loop through both loops on the hook. That is one single crochet stitch. Repeat until you reach the last of the chain stitches. The top of the row should look like a row of chain stitches. Now make one chain stitch and turn the piece around for the next row. Skip the first stitch, then single crochet in the next stitch—except from now on, slide the hook under the V, instead of into the middle of it. When you get to the end of the row, be sure to make a stitch in the last V (it's a little hard to see).

FIGURE 4-12: **To single crochet, find the second V-shaped chain stitch.**

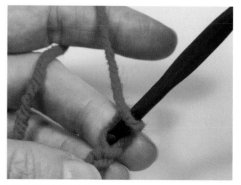

FIGURE 4-13: **Insert the hook into the middle of the second chain stitch.**

FIGURE 4-14: **Yarn over and pull the loop of yarn back through so that you have two loops on the hook.**

FIGURE 4-15: **Yarn over again and pull the new loop through both loops on the hook.**

FIGURE 4-16: Chain 1 and turn the piece around.

FIGURE 4-17: From the second row on, insert the hook under the V for each stitch.

FIGURE 4-18: The last V in the row is hard to see.

FIGURE 4-19: Swatch of single crochet

Increase (inc)

To add a stitch to a row without creating a bump on the side, make two single crochets in the same stitch somewhere in the middle of the row.

Decrease (dec)

To shorten a row by one stitch, insert the hook into one stitch, yarn over, and pull back a loop. Do the same with the next stitch. You now

FIGURE 4-20: Increase by making an extra stitch.

have three loops on the hook. Yarn over and draw the yarn through all three loops.

FIGURE 4-21: **Insert the hook through two stitches.**

FIGURE 4-22: **Draw the yarn through all three loops.**

Slip stitch (sl st)

The slip stitch can be used to make very flat stitches along a row. Many patterns also use the slip stitch to join the ends of a chain or row to make a ring. To make a slip stitch along a row, insert the hook through a stitch, yarn over, and pull the yarn through both loops on the hook. You will end up with one loop on the hook.

FIGURE 4-23: **A swatch with a row of slip stitch on top of rows of single crochet**

Change to a New Piece of Yarn

Use this technique to start a new piece of yarn if your skein runs out or you want to change colors. First, stop halfway through a stitch when you have two loops on the hook. Cut the old yarn, leaving a tail about 4 inches (10 cm) long. Take the new yarn, leaving the same length tail, and pull it through both loops. Continue as before.

FIGURE 4-24: **End the old yarn when you have two loops on the hook.**

FIGURE 4-25: **Pull the new yarn through both loops.**

Fasten Off

To securely end a row of stitches, cut the yarn, leaving a tail. Hook the yarn through the last loop, pull the tail all the way through, and pull the loop tight.

FIGURE 4-26: **Pull the tail through the last loop to fasten off.**

Weave in the Ends

Thread the tail end of the yarn onto a yarn needle and weave it through a few stitches on the back side of your project. Clip off any excess yarn so the end doesn't show. To hide a yarn tail in the middle of a piece, use the tail and the new yarn together for a few stitches.

FIGURE 4-27: **Weave the tail through a few stitches.**

PROJECT:
CROCHET HOT COCOA MUG/ ROLL-UP SCARF

⊙ MATERIALS

Worsted weight yarn in three colors: warm medium brown (like chocolate milk), whipped creamy white, and any color for your mug (stiffer yarn and tighter stitches will help the pieces stand up and keep their shape)

⊙ TOOLS

Crochet hook, size H

Yarn needle

FIGURE 4-28: **A scarf that rolls up to look like a mug of hot cocoa**

Amigurumi are little crocheted dolls in the shape of characters or objects. They were invented in Japan in the 1950s, when the craze for all things cute (or *kawaii*) began and are still popular today with people who make and design crochet projects. Rebecca Angel Maxwell (facebook.com/rebeccaangelmusic), a creativity educator and tea lover, enjoyed making amigurumi tea cups until a friend asked her, "But what do you do with it?" Then one day, her son remarked that the brown and white scarf she was working on looked like a cinnamon bun when it was rolled up. Maxwell decided to start making amigurumi with a purpose. Today she makes scarves that roll up into all kinds of drinks and food, from cupcakes to sushi.

FIGURE 4-29: **Unrolled scarf**

The directions here are an adaptation of Maxwell's design for a mug of hot chocolate that will keep your neck warm in chilly weather. Once you understand the process, you can use it to make any cylinder-shaped design you like!

NOTES:

- ☐ The bottom of the "mug" is the side where the tail of yarn hangs off the beginning chain. The last row of every section should end at the bottom as well.

- ☐ The number of rows per section is just a suggestion. Be sure to check the design by rolling up the piece as you go. Add or subtract rows as needed.

- ☐ Don't forget to make one chain stitch when you turn at the end of each row. (See the single crochet directions in the "Quick Crochet How-To" section earlier in this chapter.)

1. Start from the middle of the mug, with the whipped-cream-colored yarn. The first few pairs of rows get shorter and shorter along the top, like stairs going down, to create a pointy mound. Don't forget to chain 1 and turn at the end of every row. Here are the directions in "crochet code" form:

FIGURE 4-30: **The Whipped Cream section starts tall and gets shorter.**

- ⊙ Chain (ch) 25

- ⊙ Rows 1 and 2: Single crochet (sc) 24

- ⊙ Rows 3 and 4: sc 23

- ⊙ Rows 5 and 6: sc 22

- ⊙ Rows 7 through 10: sc 21

- ⊙ Rows 11 through 26 (or however many you need): sc 20

When you get to the last stitch of the last row of whipped cream-colored yarn, you will switch to the cocoa-colored yarn. (See "Change to a New Piece of Yarn" in "Quick Crochet How-To.")

2. The top of the cocoa section is one stitch lower than the whipped cream. For each row, single crochet 19 stitches. Repeat for 160 rows or as long as you need to. Set aside.

FIGURE 4-31: **The cocoa section is shorter than the whipped cream.**

FIGURE 4-32: **The whipped cream rolled up in the cocoa section.**

3. Now use the mug-colored yarn to make the handle separately. It's too narrow to crochet easily as a tube, so make it flat and then sew it up. To begin, tie a slip knot, leaving a 10-inch (25-cm) tail, and chain 23. Measure the chain against the side of the mug to make sure the handle will be long enough to fit your hand through and adjust as needed. Then single crochet 22 stitches (or one less stitch than the length of the chain) for six rows. Fasten off, leaving a 10-inch (25-cm) tail hanging at the opposite end of the handle from the beginning tail. Fold the rectangle up the long way to make a squarish tube. Thread a separate piece of mug-colored yarn onto the yarn needle and sew the edges together. Stuff any excess sewing yarn into the tube. Add more stuffing if you like.

FIGURE 4-33: **Crochet the handle flat.**   FIGURE 4-34: **Sew the handle into a tube.**

4. Now go back to the scarf and switch to the mug color in the last stitch of the cocoa section. Instead of making one chain stitch before you turn, make two—in crochet code, ch 2. This will make the lip of the mug stick up above the cocoa. Single crochet 20 stitches (one more stitch than the height of the cocoa) for 48 rows, or long enough for the mug color to wrap completely around the cocoa. Stop here for now, but don't cut the yarn just yet.

FIGURE 4-35: **Make an extra chain stitch before you start the mug portion to make it taller than the cocoa portion.**

5. To attach the handle, place the top end of the "tube" against the mug, one row into the mug section from the cocoa section and one stitch below the lip of the mug. Use the yarn tail you left in Step 3 and the yarn needle to sew the top end to the mug. Fasten off and weave in the yarn tail. Repeat with the bottom end, making sure that the handle is straight.

FIGURE 4-36: **Sew the handle onto the mug part of the scarf.**

FIGURE 4-37: **The finished handle**

6. Roll up the scarf to make sure the end of the mug section sits smoothly against the handle. Adjust the rows as needed. If you want the end of the scarf to be straight, fasten off. You can also make a tab that fits through the opening of the handle (where your hand goes) and hooks around the other side. The tab will hold the end of the scarf in place when it is rolled

FIGURE 4-38: **Make a tab to hold the mug end closed.**

up. To make the tab, slip stitch 3. You should be even with one end of the handle opening when the scarf is rolled up. Single crochet one row. Stop when you reach the other end of the handle opening. Make two more single stitch rows so the tab is long enough to pass through the handle opening. Then single stitch one more row, but increase 1 stitch at the beginning and 1 stitch at the end of the row. The extra stitches at each end should hook around the other side of the handle and hold the tab in place. If everything fits, fasten off.

7. Anchor and weave in any remaining loose ends. Display your scarf/ mug on a shelf, or wear it by slipping the whipped cream end through the mug handle loop.

FIGURE 4-39: **To wear, slip the whipped cream end of the scarf through the mug handle.**

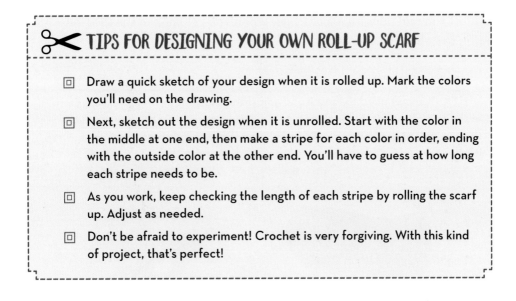

✂ **TIPS FOR DESIGNING YOUR OWN ROLL-UP SCARF**

- ☐ Draw a quick sketch of your design when it is rolled up. Mark the colors you'll need on the drawing.

- ☐ Next, sketch out the design when it is unrolled. Start with the color in the middle at one end, then make a stripe for each color in order, ending with the outside color at the other end. You'll have to guess at how long each stripe needs to be.

- ☐ As you work, keep checking the length of each stripe by rolling the scarf up. Adjust as needed.

- ☐ Don't be afraid to experiment! Crochet is very forgiving. With this kind of project, that's perfect!

QUICK KNITTING HOW-TO

There are endless variations on every knitting technique, but these basics will get you started:

Cast On

Before you knit, you have to attach the yarn to one of the needles. This is called *casting on*. The fastest method is the backward loop. Start with the needle in your right hand and the yarn on your left. (If you are left-handed, reverse directions for left and right if you are more comfortable that way.) Make a slip knot (see the "Quick Crochet How-To" earlier in the chapter) to anchor the yarn, leaving a tail. Take the working end of the yarn (the end still attached to the skein) and make a loop of yarn, with the working end of the yarn on top. (Optional: Stick your left thumb up and wrap the loop of yarn around it, as shown in Figure 4-41.) Insert the tip of the needle into the loop and slide the loop onto the needle. Pull the yarn so it is tight but not stretched. Continue to cast on as many stitches as you need and pull the yarn tight so they are touching. As you knit your first row, try to keep them tight. If you end up with excess yarn when you get back to the first loop, just pull it to tighten it.

FIGURE 4-40: **Cast on using the backward loop method**

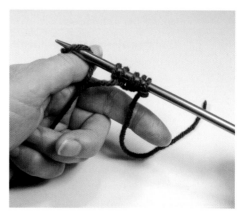

FIGURE 4-41: **Wrap the yarn around your thumb to make the loops more quickly.**

Knit

Knitting is the most basic stitch. Hold the needle with the stitches in your left hand and the yarn in the back. (If you are left-handed, reverse right and left hands if it is more comfortable.)

Insert the right needle under the left needle and through the "front leg" of first stitch (nearest the tip of the needle), from front to back. Loop the yarn over the tip of the right needle. Pull the needle back toward you, pointing it up so the tip is now over the left needle and the new stitch is on the right needle. Slide the stitch that you've just knitted into from the left needle onto the right needle. Continue with the remaining stitches. At the end of the row, turn the right needle around. It now becomes the left needle and you are ready for the next row.

FIGURE 4-42: **To do a basic stitch, insert the needle through the first stitch from front to back.**

FIGURE 4-43: **Loop the yarn over the tip of the needle.**

FIGURE 4-44: **Pull the needle back toward you.**

FIGURE 4-45: **A row of knitting**

Purl

The purl stitch is the opposite of the knit stitch. Hold the working yarn in front, and bring the right needle through the front leg of the first stitch from back to front. Loop the yarn over the right needle and pull the loop through the stitch, toward the back. Drop the stitch you just purled from the left needle. Repeat this to the end of the row.

FIGURE 4-46: **To purl, insert the needle from back to front.**

Turning

At the end of every row, all the stitches will be on the right needle. Turn the needle around so that it becomes the left needle. Take the empty needle in your right hand. You are now ready for the next row.

Garter Stitch

To do the garter stitch, you simply knit every row. This produces horizontal ribs that look the same on the right side and the wrong side of the piece. The edges of a piece done with the garter stitch lie flat.

FIGURE 4-47: **Pull the loop toward the back.**

FIGURE 4-48: **A swatch done in garter stitch**

Stockinette Stitch

The stockinette stitch is smooth on one side and bumpy on the other. To make it, you knit one row, then purl the next. A piece made with the stockinette stitch tends to curl at the edges. You can use the curled edges as part of your design, like the t-shirts fringes you curled up in Chapter 2.

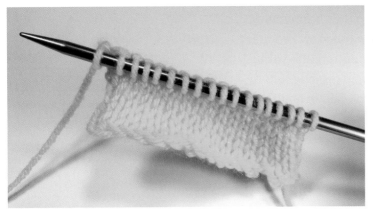

FIGURE 4-49: **A swatch done in stockinette stitch**

Ribbing

You can create vertical ribs by knitting one or more stitches, then purling one or more stitches, all along the row. In the next row, you reverse the knit and purl stitches, so the smooth and bumpy stitches line up in columns.

FIGURE 4-50: **A swatch of ribbing**

Cast Off

To finish a piece of knitting, knit two stitches onto the second needle. Then pull the first stitch over the second stitch, so only one stitch is left on the second needle. Knit another stitch onto the second needle, and repeat until the row is done. Then fasten off by cutting the yarn, leaving a tail of about 8 inches (20 cm), and pulling the tail through the last loop.

FIGURE 4-51: **To cast off, start by knitting two stitches.**

FIGURE 4-52: **Pull the first stitch over the second stitch.**

FIGURE 4-53: **Fasten off the last stitch by pulling the tail through it.**

PROJECT:
TINY TOOTHPICK KNITTING

FIGURE 4-54: A miniature ball of yarn makes the tiny piece of knitting look more real.

If you love miniatures, this tiny knitting project is for you! Wear it, hang it in your home, or give it as a gift to your favorite knitter.

1. First, prepare the toothpick/knitting needles. Snip off the points with a wire cutter or scissors and smooth them with sandpaper or a nail file. If you want to make your needles shorter, clip the other ends with wire cutters or scissors and sand them smooth. To help keep your knitting from sliding off, glue beads to the ends, like the caps on real needles.

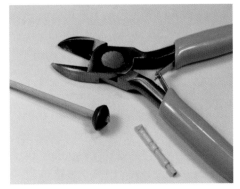

FIGURE 4-55: Make the toothpick "needles."

2. Cast on between 8 and 12 stitches, or as many as the toothpick can hold without the yarn slipping off the end.

3. Stockinette stitch (knit a row, then purl a row) for as many rows as you choose. End in the middle of a row, leaving an equal number of stitches on each needle.

4. Cut off the yarn, leaving about 4 feet (1.25 m) on the end. Roll the yarn neatly into a little ball, leaving a little loose yarn between the ball and the knitting. With the smooth side up, place the ball in the center of the knitting. Poke the needles through the back of the ball at an angle so they cross inside the ball and the points stick out the bottom. Try to catch the yarn in the needles so it doesn't unravel. If you need to, dab a little glue on the ball to keep it in place. Let it dry.

5. To wear your knitting like a pin, put a safety pin through the back of it and attach it to your shirt. To hang it on the refrigerator or message board, attach a peel-and-stick magnet to the back. (Add some drops of fabric glue if you need to.) You can also attach a loop of thread to the back and hang it in a window or as a tree ornament.

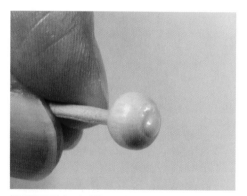

FIGURE 4-56: **Glue a bead on the end.**

FIGURE 4-57: **Cast on.**

FIGURE 4-58: **Cast on a row.**

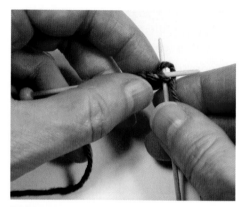

FIGURE 4-59: **Knit as usual.**

FIGURE 4-60: Roll the tail into a mini ball of yarn.

FIGURE 4-61: Poke the needles through the ball of yarn to hold them in place.

PROJECT:
SPOOL KNITTER

◉ MATERIALS

Wooden spool, about 1½ inch (3.8 cm) across (or wider—found with other wood shapes in craft stores)

4 finishing nails, about 1 inch (2.5 cm) long (check to see that they are smooth with nothing sticking out to catch on the yarn)

Short wooden dowel, ⅛ inch (3 mm) wide and 3½ inches (9 cm) long

(Optional) Wooden bead, large enough to fit snugly onto dowel, and glue

Yarn (thinner is better)

◉ TOOLS

Hammer

Pencil sharpener

Sandpaper

FIGURE 4-62: **A spool knitter**

✂ FABRIC AND FIBER INVENTORS: ELIZABETH ZIMMERMAN

Elizabeth Zimmerman, known as "the grandmother of modern knitting," began writing about the art form in 1955 and produced many books, workshops, TV series, and videos before her death in 1999. She called the methods and designs she created "unventions" because she believed other people had probably invented them before she did. Among the things Zimmerman "unvented" is I-cord, a long, thin tube made with knitting needles but similar to the cord created with the spool knitter.

Zimmerman is probably best known for her method of knitting on round needles. Instead of putting together sweaters and other items from flat pieces, she knitted them in round tubes, which made them easier to shape. Her formula for figuring out how to make a tube-shaped pattern in any size, the Elizabeth Percentage System, or EPS, is still used today.

Many of Zimmerman's designs were inspired by math, such as the Pi Shawl, based on the number pi (3.141592), which represents the relationship between the outside circumference and the width of a circle. Her Baby Surprise Jacket is knitted flat and folds up like origami. And she created the Mobius scarf, also known as the infinity scarf, a flat loop with a twist in it, which gives it one continuous side.

For Zimmerman, knitting was one of the highest achievements in human history. She inspired generations of knitters to use their brains and their imagination to experiment, play around with patterns, and produce amazing designs of their own.

Also called a French knitter or a knitting nancy, this little circular loom has been a favorite with beginning knitters for generations. The long knitted tube it creates is similar to a knitting staple called "I-cord," and it has many uses. Plus it's fun and relaxing to make!

1. To make the loom, first make sure the spool is smooth, without any wood slivers that would catch on the yarn. If the hole in the center isn't smooth, roll a piece of sandpaper into a tube and stick it into the hole. Rub it around until the hole is as smooth as possible.

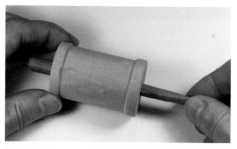

FIGURE 4-63: **Smooth the hole with a rolled-up piece of sandpaper.**

2. Next, take the spool and mark four dots on it, spaced evenly around the hole. They should be far enough apart to fit your finger in between them easily. Hammer a nail into each dot. Make sure the nails are straight and hammered in deeply enough that they don't move around. These are your pegs.

FIGURE 4-64: **Hammer in the nails.**

3. The pick, which is like a short knitting needle, is used to move the yarn around. To make it, take the dowel and sharpen one end with the pencil sharpener. Stop before it gets too sharp. Use the sandpaper to smooth the tip to a rounded point. (Optional) Glue the bead on the other end of the dowel and let it dry.

FIGURE 4-65: **Nails in the spool**

FIGURE 4-66: **Sharpen a dowel to make a rounded point on the pick.**

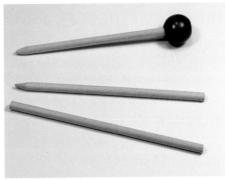

FIGURE 4-67: **Glue a bead on the pick for a better grip.**

4. To use the loom, take the yarn and tie a knot in the end. Drop the knotted end into the hole in the center of the spool until it comes out the bottom.

5. Hold the tail fairly tight with your left thumb (or if you're left-handed, with your right thumb). Wrap the yarn once around the nearest nail in a counterclockwise direction. Then bring the yarn around to the next nail going counterclockwise around the spool. Repeat, wrapping it around that nail in the same direction. Repeat with the remaining nails.

FIGURE 4-68: **Drop the yarn through the hole.**

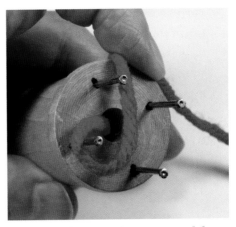

FIGURE 4-69: **Wrap the yarn around the first nail.**

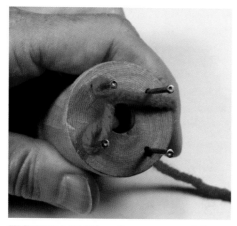

FIGURE 4-70: **Wrap the yarn around the second nail in the same direction.**

FIGURE 4-71: **End the first round at the first nail.**

6. When you get back to the first nail, make a second round, but this time just circle all the nails with the yarn in one big loop, above the first round of little loops.

7. Now you will make the first stitch. On the nail you started with, use the pick to lift the bottom loop over the straight yarn above it and right over the top of the nail. Repeat with the other nails. When you finish the round, tug on the tail of the yarn a little to pull the knitting into the hole.

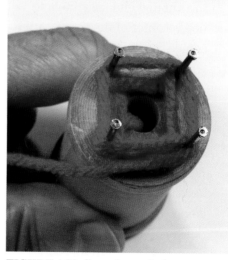

FIGURE 4-72: **Second round of yarn**

8. Repeat Steps 6 and 7 as many times as you like. The knitted tube you are making will start to come out of the hole at the bottom.

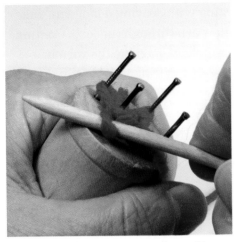

FIGURE 4-73: Lift the bottom loop with the pick.

FIGURE 4-74: Bring the bottom loop over the top yarn on the nail.

FIGURE 4-75: The knitted cord comes out the bottom of the spool.

9. To cast off, cut the yarn you are working with, leaving a long tail. Take a stitch as usual, but pull the tail of the yarn through the loop. Continue with the remaining stitches. To leave the end open, tie the yarn in a knot. To close it, pull the yarn tight before knotting it. Weave in the end.

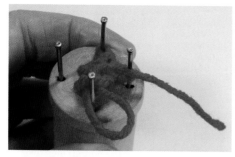

FIGURE 4-76: **Cast off by pulling the end of the yarn through the loop.**

FIGURE 4-77: **Pull the tail to close up the end.**

Adaptations/Extensions:

- ▣ To make wider cord, use a cardboard toilet paper tube and tape craft sticks around it as pegs. Secure them with colorful duct tape. This version is good for younger kids.

- ▣ Finger knitting is done the same way as spool knitting, but it produces a thin flat piece instead of a tube. To cast on, hold the tail of the yarn against your palm with your thumb, and wind the yarn in and out of your fingers. Then go the other way. This is your first row of "loops." Repeat to make a second row of loops. Lift the bottom loop over the top loop on each finger, just like with the spool knitter. For the next row, just bring the working yarn across the front of all your fingers instead of weaving it in and out. Continue for as many rows as you like. To cast off, stop when you have one loop on each finger. Bring the loop from your first finger over the loop on your second finger. Bring the bottom loop over the top loop as usual. Repeat with your other fingers until you have only one loop left. Cut the yarn and pull the tail through the remaining loop. Pull it tight to knot it.

- ▣ Use your Spool Knitter to make a stretch sensor out of conductive yarn! (See Chapter 5.)

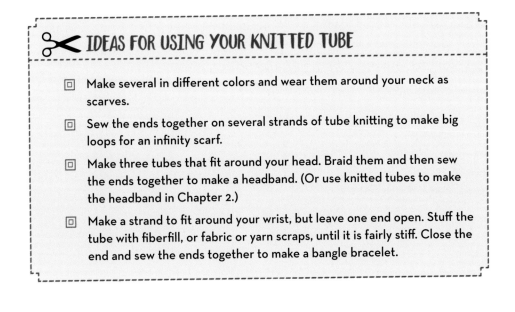

✂ IDEAS FOR USING YOUR KNITTED TUBE

- ☐ Make several in different colors and wear them around your neck as scarves.

- ☐ Sew the ends together on several strands of tube knitting to make big loops for an infinity scarf.

- ☐ Make three tubes that fit around your head. Braid them and then sew the ends together to make a headband. (Or use knitted tubes to make the headband in Chapter 2.)

- ☐ Make a strand to fit around your wrist, but leave one end open. Stuff the tube with fiberfill, or fabric or yarn scraps, until it is fairly stiff. Close the end and sew the ends together to make a bangle bracelet.

SOFT TECH

Electrify your fabric and fiber designs with lights, motors, sensors, and more!

✂ FABRIC AND FIBER INVENTORS: LEAH BUECHLEY

The invention of the LilyPad Arduino toolkit by computer scientist and artist Leah Buechley helped launch the age of e-textiles. In 2006, Buechley combined the artist-friendly Arduino micro-controller (a small programmable circuit board that can control lights, motors, and sensors) with her own sewable electronic components to bridge the gap between technology and textile design. Between 2009 and 2014, as head of the High-Low Tech group at the MIT Media Lab, she helped spread her ideas and techniques to students, engineers, artists, fashion designers, and crafters everywhere. Many of the most active innovators in the field of soft circuits today got their start studying under Buechley at MIT.

FIGURE 5-1: A circuit of sewable LilyPad components

"I designed the LilyPad to help people make soft, beautiful, wearable technology," Buechley says. "I hope it's been fun for people to play with. I sure love seeing all of the wonderful things people are building with it!"

FIGURE 5-2: Leah Buechley. Credit: MIT Media Lab.

Electronic devices are getting smaller all the time. Meanwhile, the materials to hold them together have gotten lighter and stronger. One result has been the invention of *e-textiles*—fabric and fiber designs that include electronic components or parts.

Makers are using these new kinds of materials to create "smart clothing" that can sense what's going on in your body or the environment around you and send the information to a computer or other device. For example, a smart workout shirt can measure your heart rate as you exercise. Running shoes can tell you the distance you've run using a GPS. A smart bike jacket can flash turn signal lights when you're riding in traffic. And smart hats or socks for infants can tell parents if their baby is breathing properly at naptime. At places like the Georgia Institute of Technology, researchers are even trying to create fabric that can make its own electricity using solar power, wind energy, and your own movements.

In this chapter, you'll learn how to create interesting effects using only a few basic electronic components and some clever design. Use these ideas to add soft circuits to any kind of fabric or fiber invention!

HOW TO MAKE AN ELECTRICAL CIRCUIT

Before you can add electronics to your fabric and fiber inventions, it helps to understand a little bit about how electricity works. Here are some things you need to know:

Electricity is the flow of charged particles.
Every substance in the universe is made up of ultra-tiny *atoms*. Atoms contain three kinds of particles: *protons* and *neutrons* in the center (called the nucleus), and *electrons* that circle around the nucleus in a cloud. Protons carry a positive (+) charge. Electrons carry a negative (–) charge. (Neutrons have no charge—they're neutral.) Electricity is the movement of electrons between atoms.

Conductive materials are good at carrying an electrical charge.
In a good conductor, such as most metals, electrons can jump from one atom to another very easily. Materials that are less conductive—such as graphite (used to make pencil lead), water, and your skin—can also be used for some projects.

Materials that do not conduct electricity very well are called _insulators_.
Good insulators include fabric and fibers like wool and cotton, rubber and plastic, paper and cardboard, and air. Many types of wire are coated with insulation. If you want to use them in your projects, strip the insulation off the ends using a wire stripper or sandpaper to reveal the conductive inner core.

An _electrical circuit_ is a path for electricity to travel along.
To make a circuit, you need to surround the conductive path with insulating material. This keeps the electricity where you want it.

A circuit must be _closed_ for electricity to flow.
Electrons jump when there's another atom nearby ready to receive them. To make sure there's always another place for the electron to jump to, the ends of a circuit connect to form a loop. When the circuit is closed, the electrons on the last atom on one end of the circuit can jump

to the first atom on the other end and keep going around. If a gap in the circuit keeps the electrons from moving, the circuit is *open*.

A circuit has three measurements: current, voltage, and resistance. The *voltage* (which is measured in volts) tells you how much electrical charge is available in the circuit to do work. The *current* (which is measured in amps) shows how strong the flow of electricity is by telling you how many electrons are traveling through the circuit per second. The *resistance* (measured in ohms) tells you how easily current can pass through the circuit. All three measurements are related. As the resistance goes up, the current goes down, and if you multiply them together, you get the voltage. (This is known as *Ohm's Law*: voltage equals current times resistance.) Components called *resistors* are used to reduce the flow of electricity through the circuit so it doesn't overload sensitive parts. In soft circuits, you can use materials with a higher resistance to limit the current the same way a resistor does in a regular circuit.

✂ WORKING WITH COIN BATTERIES

The power for most soft circuits comes from batteries. Batteries have a negative (–) end, or terminal, and a positive (+) terminal. The batteries that you will use for the projects in this book are 3 volt (3V) coin batteries, which are used in watches and other small devices. They are lightweight and won't cause an electric shock if you happen to touch the circuit!

On a coin battery, one side is positive and the other side is negative. You can buy or make a battery holder to connect each end of the circuit to the proper side of the battery. Make sure the two sides do not connect with each other to cause a short circuit!

FIGURE 5-3: The positive and negative sides of a 3-volt coin battery

The components that are powered by a circuit are called the *load*.
A load can be a light, a motor, or a speaker. It can also be a sensor that collects information or sends it to a device. The main type of load that you will use for the projects in this book are LEDs (light-emitting diodes), a type of light bulb that needs only a tiny amount of electricity to work.

All electrical components have positive and negative ends.
For some kinds of components, it doesn't matter how you connect them to the circuit. Others will work only if the negative end of the component is connected to the part of the circuit coming from the negative terminal of the battery (and the positive end is connected to the part of the circuit going to the positive terminal of the battery). Components like this are said to have *polarity*. To help you keep track of which side of your circuit is positive and which is negative, you may want to mark (–) and (+) along the path.

Components in a circuit can be connected in two different ways: in series or in parallel.
When components are in *series*, the current must flow through one component to get to the next. It takes more voltage (meaning more batteries) to power components that are wired in series. With a *parallel* circuit, each component has its own pathway to the battery. The current is reduced, but the voltage stays the same. For the projects in this book, be sure to connect your LEDs in parallel.

FIGURE 5-4: LEDs that are connected in series

FIGURE 5-5: LEDs that are connected in parallel

A *circuit board* is a piece of nonconductive material that holds the circuit and some or all of its components.

The lines on a circuit board that carry electricity between the components are called the *traces*. On regular circuit boards, the traces are metallic. Circuit boards also have conductive pads where components can be added using solder (melted metal that hardens like glue). Soft circuit boards are made of nonconductive fabric, felt, or other lightweight material. They use traces and pads made of conductive fabric, fiber, wire, ink, or other flexible substances.

A *short circuit* occurs when two parts of a circuit connect where they're not supposed to.

A short circuit can keep your circuit from working properly. It can also be dangerous if it prevents the electricity from reaching the load. The traces or wires can overheat and fry components or even cause a fire.

A *switch* controls the flow of electricity.

When you turn on a switch, you are closing a circuit. Think of it like a drawbridge. When the switch is open, it stops the flow of electricity. When the switch is closed, it lets electrons travel over a gap.

A *sensor* is like a switch that works differently under different conditions.

Some sensors are designed to physically open or close a circuit, like a tilt sensor. Others are made using *variable resistors*, like a stretch sensor. Variable resistors are materials that can carry a higher or lower amount of current, depending on certain conditions.

QUICK SOFT CIRCUIT HOW-TO

The circuits in this chapter use regular yarn, felt, or other fabric, and the same kind of knitting and hand-sewing tools you used in earlier projects. You'll also need some additional sewing and crafting materials, as well as a few special supplies including the following:

- ☐ Conductive thread (any type will do)
- ☐ Conductive yarn made of steel and polyester (for the stretch sensor)
- ☐ LEDs (preferably the round, 3-mm size with long wire leads—red and yellow are the brightest)
- ☐ 3V (volt) coin batteries (CR2032 or similar sizes)
- ☐ Metal and regular small beads
- ☐ (Optional) Sew-on metal snaps
- ☐ Needle-nosed and/or round pliers

To find out where to get these materials, see "The Fabric and Fiber Supply Basket" section at the front of the book.

Working with Conductive Thread

Conductive thread is made by twisting thin strands of metal (such as silver or steel) together or by combining them with other kinds of fiber, like nylon or polyester. It's a little bit harder to work with than regular thread because it tends to unravel. When the end of the thread splits into a bunch of loose strands, it can be harder to fit it through the eye of a needle. Since it's not insulated like wire, you need to be sure that loose strands or stray stitches don't touch other places on the thread or any components and cause a short circuit. And you may find that conductive thread can untie itself from knots or twist itself into tangles. To avoid these problems, here are some suggestions:

- ☐ To make it easier to get conductive thread on the needle, use needles with large eyes. A needle threader is also helpful. (See needle threading tips in the Introduction at the front of the book.)
- ☐ Knot the thread at the beginning of every row of stitches. For extra security, anchor the thread by making a few small stitches in one place.

Then, as you start to sew with the thread, wrap the first few stitches around the tail so it doesn't hang loose.

- At the end of a row of stitching, knot the thread around existing stitches from the same piece of thread. Then weave the tail through a few stitches and trim it so the end doesn't show. To avoid tangles, go slowly and try to keep the thread smooth as you work. If it gets twisted up, hold your sewing by the fabric (or the needle) so the thread spins and untangles itself.

- You can combine conductive thread with ordinary thread or yarn for sewing, knitting, weaving, and crochet projects. This helps hold it in place and adds a little insulation between neighboring stitches.

- To avoid a broken electrical connection, try to sew each trace with a single piece of conductive thread. If you need to connect two pieces of thread, tie the new piece around a stitch in the old piece.

FIGURE 5-6: Conductive thread on a small bobbin (front left) and a large spool (back) and conductive yarn (front right)

- Avoid designing circuits with traces that touch or cross whenever possible. When you must cross rows of stitching, cover the bottom row with one or more layers of insulating fabric. Then sew the top row in this bridge—making sure the two rows don't touch.

- When you are done sewing, you can seal and insulate knots in conductive thread with a dab of hot glue or clear nail polish.

FIGURE 5-7: Weaving a loose tail under existing stitches

- Here are other ways to insulate conductive thread so it doesn't short-circuit:

  - ☉ Use a thick piece of felt, or use iron-on adhesive to stick two or more layers of felt together so you can sew a row on one side that doesn't go all the way through to the other side. Neoprene fabric (the stuff that mouse pads and foam sleeves for soda cans are made from) is also good for sewing thick soft circuit projects.

  - ☉ Insert the conductive thread into piping or another kind of fabric trim that contains hollow "tunnels."

  - ☉ Cover the thread with stuffing or batting (the same material you use for a stuffed animal or a quilt).

  - ☉ Coat the thread with puffy fabric paint, hot glue, or clear nail polish.

  - ☉ Wrap the thread with stitches using any nonconductive yarn or thread.

FIGURE 5-8: **Sealing a knot with hot glue**

Other Conductive Materials for Traces

There are other specialized and ordinary materials you can use for circuits, such as these:

Fabric adhesive tape
Peel-and-stick tape that works with felt and other fabrics.

Conductive fabrics
Useful for cutting out patches and pads.

Conductive ribbon
Use for conductive trim instead of thread.

FIGURE 5-9: **Samples of conductive fabric, ribbon, and adhesive tape. The black velostatic plastic becomes more conductive under pressure.**

Soft jewelry wire, floral wire, or magnet wire

If the wire is covered or coated with insulation, use a wire stripper or a piece of sandpaper to remove it.

Copper pot scrubber

Pull out the "knitted" stitches in the scrubber and unwind the flat copper wire. It's very conductive but not insulated. Use it for weaving (making sure to separate the rows with regular yarn to avoid short circuits) or sew it to the head of a stuffed fabric creature to make conductive curly hair.

Aluminum foil

Crumple it up into a small ball to make a conductive bead, or iron it onto a sheet of adhesive interfacing (which only has glue on one side) to use it like fabric.

FIGURE 5-10: **Conductive jewelry and electrical wire**

FIGURE 5-11: **Copper pot scrubber**

Making Conductive Pads

Conductive pads give you a place to connect additional electronic components. Here are some options for how to make them:

- Use conductive thread to sew a small, smooth patch onto a piece of fabric using a satin stitch. (See "It's Sew Basic" at the front of the book.) For a conductive dot, make a star of crisscrossing stitches. Fill a larger area with a field of small Xs, spacing them around in any pattern.

- Use a sewable metal snap, which consists of two metal disks: one is a stud that sticks up, the other is a socket that fits over it. The stud is sewn facing away from the fabric, and the socket is sewn facing down, into the fabric. Sew snaps near the edge of a piece of fabric if you want to attach alligator clip wires to them. Or sew matching snaps onto another piece of fabric to attach it. To use snaps as a switch, sew a stud and socket pair onto one piece of fabric, far enough apart that you can fold the fabric over and snap them together. Or make a soft "wire" by sewing a row of conductive stitches along a strip of fabric and placing snaps on the ends.

▣ Try other materials, such as a piece of conductive fabric, or metallic sewing supplies like hooks and eyes, magnetic clasps, metal beads, buttons, or brads.

FIGURE 5-12: **Conductive metal buttons (top left), snaps (top right), and hooks (bottom left)**

FIGURE 5-13: **Conductive metal beads and bells**

Working with LEDs

E-textiles use LED lights because they're small, can work with low-voltage batteries, don't get very hot, and last (almost) forever. Here's what you need to know about LEDs in soft circuits:

▣ LED stands for *light-emitting diode*. A *diode* is a component that lets current flow in only one direction. So to work, the positive side of an LED must be connected to the positive side of

FIGURE 5-14: **Testing LEDs together on a battery**

the battery and the negative side of the LED to the negative side of the battery. On standard LEDs with a round bulb (called the case) and two wires (the leads), the positive lead is usually longer than the negative lead. The negative side of the case may also be slightly flattened.

☐ To test your LED, slide the battery between the leads. If it doesn't light up, flip the battery around.

☐ To prepare your LED for sewing, bend the leads so the top points forward, toward you. That's the brightest part of the LED. Then use round and/or needle-nosed jewelry pliers to shape the leads into loops or spirals. A handy trick to help you tell them apart is to make the positive lead into a round spiral and the negative lead into a square or triangular spiral.

☐ If you are putting more than one LED in a circuit, wire them in parallel. (See the "Sew a Soft Circuit Tester" project for directions.)

☐ Different colors and kinds of LEDs take different voltages. If the voltage is too low, they may shine only dimly or not at all. If

FIGURE 5-15: **Circuit Sticker LEDs (top left), round LEDs (left to right, 3-mm, 5-mm, and 10-mm sizes), and LilyPad LEDs (bottom)**

FIGURE 5-16: **Curling an LED wire lead with round pliers**

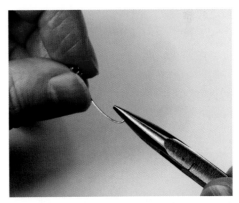

FIGURE 5-17: **Bending the lead into a round spiral with needle-nosed pliers**

it's too high, you can fry them. Most will work with a 3V battery without danger of overheating.

☐ LEDs with different voltages may not work together in the same circuit. Test them by sliding the ones you want to use together over a battery to see if they all light up (see Figure 5-14). Some types of sewable LEDs, such as LilyPad or Chibitronics Circuit Stickers, have on-board resistors, so you can combine LEDs with different voltages in the same circuit.

FIGURE 5-18: **An LED ready for sewing. The round side is positive.**

Other Kinds of Loads

Other output devices that work with your 3V battery can be built into your soft circuit. They include vibration motors (tiny motors that shake when activated) and buzzers (which produce squeals and squeaks depending on their voltage). Speakers with their own batteries, like those on greeting cards, also work in simple soft circuits.

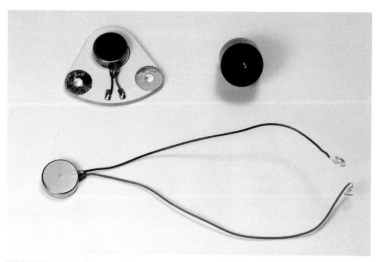

FIGURE 5-19: **A sewable motionboard from Teknikio (**teknikio.com**, top left), and standard motors and buzzers that work with simple e-textile projects**

Adding Control

If you'd like to go beyond the projects in this chapter, that are entry-level sewable microcontrollers that let you program lights and other outputs to react to homemade or manufactured sensors. Along with LilyPad Arduino, boards to check out include these:

◻ Gemma is a stripped-down Arduino board from Adafruit (adafruit.com), a New York–based electronics manufacturer, founded by engineer Limor Fried in her MIT dorm room in 2005. The stuffed burlap FiberBot project in my book *Making Simple Robots* (Maker Media 2014) uses an Adafruit Gemma and LED matrix (grid of lights) to create an animated eye.

◻ Chibitronics Circuit Stickers is a system of peel-and-stick LEDs invented by Jie Qi when she was a student under Leah Buechley at MIT's High-Low Tech research group. These kits include special effects modules that make your LEDs flash in several different patterns. Created for use on paper, they can also be sewn into e-textile projects. Chibitronics has also introduced a programmable controller called Love to Code.

◻ Micro:bit is a mini-computer developed by Microsoft for the BBC. It is small enough to be built into wearable projects and includes an onboard light matrix, sensors, and Bluetooth so you can send and receive information to the micro:bit by phone or tablet.

FIGURE 5-20: **Three types of programmable microcontrollers: a Microsoft micro:bit (top left), a ProtoSnap LilyPad Development Board (right), and an Adafruit Gemma (bottom left)**

Working with Felt

Felt is great for soft circuits because it's thick, easy to sew, and doesn't fray when you cut it. Here are some tips for using it in your projects:

☐ Regular craft felt comes in sheets about as big as letter-sized paper. Avoid felt with sparkles or other additives.

☐ Peel-and-stick felt sheets are a little stiffer. Use them to cover the back of your circuit to keep the stitches from showing (and to prevent short circuits if they touch something conductive). You can sew through these sheets, but the glue may gum up your needle.

FIGURE 5-21: Regular, peel-and-stick, and thick felt

☐ Thicker stiffened felt may be easier to find precut into large shapes (such as butterflies or flowers), but you can cut several smaller pieces out of them pretty easily. This felt is about ¼-inch (6-mm) thick and makes a good base for many projects. You may need a thimble to help push a needle through it.

☐ You can also attach a second layer of regular felt to your base using iron-on adhesive. If you do this before you sew on your LEDs or other components, you can use the layers to separate rows of stitching on the back and the front side.

☐ For projects that don't need as much stiffness, you can cut up felted wool sweaters to make your own felt shapes (see Chapter 2).

SAFETY WARNING:

☐ If any part of your circuit becomes hot or burns out, disconnect the battery immediately. Check for any short circuits before turning it back on.

☐ When cutting or stripping wires, snipped bits may go flying. Protect your eyes by wearing safety glasses.

✂ FABRIC AND FIBER INVENTORS: HANNAH PERNER-WILSON AND MIKA SATOMI

The projects in this chapter are adapted from the *open source* (free to use) inventions of KOBAKANT, a collective formed in 2006 by European based e-textile designers Hannah Perner-Wilson (www.plusea.at/) and Mika Satomi. Their website, KOBAKANT/HOW TO GET WHAT YOU WANT (kobakant.at/DIY), is one of the best places to find information on making sewable sensors, switches, and other electronic components using crafts supplies and soft materials.

Perner-Wilson says she learned by studying other people's examples from the Internet, so she feels it is natural to give back to the community the same way.

But *documenting* (recording what she does) when she makes a project isn't just about providing step-by-step instructions. "It is also a means of telling the story of my process and promoting a hands-on, exploratory approach to working in this field," she says. "I aim to capture and share not only what works, and the polished final designs, but also the attempts and failures along the way. Understanding that failure is part of a successful process is one of the first things you need to learn, so that you don't give up too quickly."

FIGURE 5-22: Hannah Perner-Wilson wearing a vest that holds her e-textile sewing tools (CC BY 2.0)

A graduate of Leah Buechley's High-Low Tech research group at MIT, Perner-Wilson has also explored techniques for carving traces, painting speakers, casting pixels, and sculpting motors and other handmade electronic components as part of a "Kit of No Parts." (She also helped design and build the mi.mu electronic musical glove invented by singer Imogen Heap featured in my book *Musical Inventions* [Maker Media 2017]. After returning from an expedition to the African island nation of Madagascar in 2015, she began working on a "wearable studio" with portable tools and equipment. "Carrying my tools with me allows me to practice my trade in all kinds of environments and situations," Perner-Wilson says. "It also means that I open my practice to a world outside my studio walls, exposing what I do and inviting others in."

In 2017, KOBAKANT decided to create a shop to make custom-designed e-textiles. The store, located in Berlin, Germany, is called KOBA Maßschneiderei (a combination of Japanese and German words meaning "family-run tailor shop"). The patterns they design are still open source, but any items they make belong to the people who pay for them. "We hope to find out what people want e-textile and wearable technology to do for them by providing a place to come with their ideas," Perner-Wilson explains. "What if this shop were opening in your neighborhood? Could you imagine coming in to place an order for something? What would it be?"

When she started, Perner-Wilson liked the idea of using sewing and crafts in unusual ways. Her biggest problems were finding sources of interesting conductive materials, and figuring out how to make soft, flexible, reliable, wearable, comfortable electrical sensors and connections using textiles.

Now, she thinks that "the hardest problem for makers to solve is, how do we balance our love for, and impact on, the world? Maybe we should think of ourselves not as makers, but as explorers. Every time we discover a new material in a local market or on the Internet, we must examine it closely, gather information about it, and decide if this is something want to support through our purchase and use. It is important to stay curious, playful, and experimental, but we should use these strengths not just to imagine incredible things, but also to promote ethical and environmental standards."

PROJECT:

SEW A SOFT CIRCUIT TESTER

◉ MATERIALS

2 pieces of felt—1 regular and 1 peel-and-stick—each about 3 by 4 inches (7.5 by 10 cm)

3V coin battery, CR2032 or similar

Conductive thread

Thread or embroidery floss

1–3 LEDs

(Optional) sewable metal snaps

◉ TOOLS

Fabric chalk, fabric pencil, or regular pencil

Round and needle-nosed jewelry pliers

2 hand sewing needles to fit the regular and conductive thread

Needle threader

(Optional) thimble

Straight or safety pins

2 alligator wires

How can you tell if a fabric or fiber you want to use can conduct electricity? Build a pocket-sized tester you can take with you when you go shopping

FIGURE 5.23: A tester with one LED (left) and a three LED "bar graph" (right). The more current passes through a variable resistor, the more LEDs light up on the bar graph.

for materials! Clip alligator wires onto the test pads and you can see if your mystery material closes the circuit. This tester is also good for trying out new sensor designs, like the ones in the next project.

> **REMINDER:** Be sure to knot the conductive thread and weave all loose ends into the stitching to avoid short circuits. Also, don't let two rows of stitches touch or cross.

1. Curl up the leads of the LED as described in the earlier section "Working with LEDs."

2. Draw the circuit on a piece of felt (or sketch it on a piece of paper as a guide) in the following order:

   A. Trace around the battery.

   B. Trace around the LED. Draw a line from the battery to the negative lead.

   C. Draw a line from the positive lead of the LED to the first test pad.

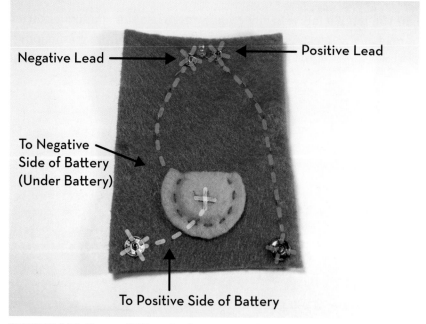

FIGURE 5-24: **For one LED, make three traces with conductive thread as shown.**

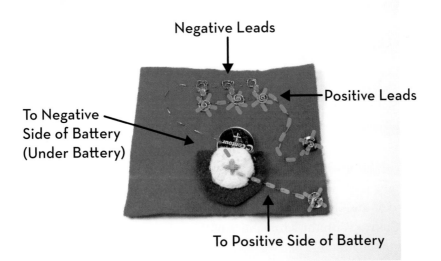

Negative Leads

Positive Leads

To Negative
Side of Battery
(Under Battery)

To Positive Side of Battery

FIGURE 5-25: **For more than one LED, make parallel traces as shown.**

D. Optional: To add more LEDs, extend the line from the battery to the negative lead of the first LED so it touches the negative leads of the other LEDs. End the line with the last LED. Then draw another line that touches all the positive leads before it goes to the first test pad.

E. Draw a line from the battery to the second test pad. Make sure it doesn't touch or cross any other lines.

3. Sew a conductive thread pad where the battery will go. With the same thread, sew a row of running stitches from the battery to the negative lead of the LED. Whip-stitch around the negative spiral with the conductive thread to attach the LED securely to the felt. (If you are using more than one LED, continue the row of stitches to connect the other negative leads.)

FIGURE 5-26: **Sew one trace from the battery to the LED.**

FIGURE 5-27: **Whip-stitch around the wire leads.**

FIGURE 5-28: **Three LEDS connected in parallel**

4. Take a new piece of conductive thread and whip-stitch around the positive lead of the LED (or LEDs) the same way you did for the negative lead in Step 3. With the same piece of thread, sew a row of running stitches to the first test pad. Sew on a snap or make a thread pad.

FIGURE 5-29: **Sew a second trace from the LED to a snap.**

FIGURE 5-30: **The back of the stitching shows the gap where the LED is connected to the circuit.**

5. To make a battery pouch, take a piece of felt, place the battery on it, and trace around the battery. Draw a larger circle around the first and cut it out. Then cut the top off to make an opening that is large enough to slide the battery in and out. Sew a conductive pad on the battery pouch in the shape of a plus sign (+) to remind you to put the battery in with the positive side facing out. Then sew running stitches to the edge of the piece of felt. Leave the rest of the conductive thread hanging off for now.

6. Use embroidery floss to sew the pocket to the base over the negative conductive pad for the battery. Be careful not to get tangled up with the loose conductive thread. Insert the battery into the pocket to make sure it fits snugly. Test it by touching the loose piece of thread to the negative lead of the LED. If it lights up, your connection is good! Remove the battery. Take the loose conductive thread on the pouch and sew a running stitch to the second test pad or snap.

7. Insert the battery into the pocket again and snap the test pads together. If the LED lights

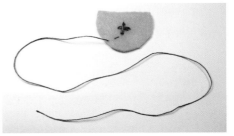

FIGURE 5-31: Sew a conductive pad and a row of stitches on the pouch, then leave the thread hanging while you do the next step.

FIGURE 5-32: Attach the pouch to the body of the circuit tester by sewing around the edge of the circle with nonconductive thread or floss. Make sure the battery fits inside snugly.

FIGURE 5-33: Use the loose end of the conductive thread on the pouch to sew a line to the second snap or pad.

FIGURE 5-34: The stitches from the back showing the battery pouch.

up, your tester is complete. If you wish, cover the back of the tester with a piece of peel-and-stick felt. Press it on to make a good seal.

FIGURE 5-35: **Connect the snaps to the tester.**

FIGURE 5-36: **Attach peel-and-stick felt backing for protection and support.**

8. To use the tester, lay a piece of material between the test pads, or use alligator wires to connect them to a sensor, switch, or anything conductive.

FIGURE 5-37: **Testing a piece of metallic fabric trim**

PROJECT:
SOFT SENSORS

◉ MATERIALS

1 thick piece or 2 small pieces of felt for each sensor, about 3 or 4 inches (8 or 10 cm) square

Conductive thread

Embroidery floss (separate it so there are only three strands) or thread

Safety pins or straight pins

Sewable metal snaps or other conductive pads or connectors (see the section "Making Conductive Pads" in the Quick Soft Circuit How-To earlier in the chapter)

◉ TOOLS

Pencil or fabric marker

2 hand sewing needles, one with an eye big enough for conductive thread

Needle threader (or thread or paper strip to make one)

(Optional) thimble

◉ ADDITIONAL MATERIALS AND TOOLS

For the Quilted Pressure Sensor:

    2 pieces of aida, canvas, or other stiff cloth, about 2.5 inches (6 cm) square

    Thick stiffened felt (or several pieces layered together with iron-on adhesive), about 2 inches square

    Quilt batting, slightly smaller than the felt squares

    Additional felt pieces

For the Tilt Sensor:

    5–10 small beads, any kind

    6 small conductive beads

    1 or more large beads, preferably conductive

For the Stretch Sensor:

    Spool Knitter (see Chapter 4 for how to make and use one)

    Steel/polyester blend conductive yarn (available from lessemf.com, catalog #306; one bobbin is enough for several sensors)

Soft sensors are made of the same material as soft circuits—no fancy electronics needed! These prototypes are designed to work with the Soft Circuit Tester, but you can also build them into any soft circuit project.

I. The Quilted Pressure Sensor

This pressure sensor is basically a push button (also known as a *momentary* switch). It closes a circuit when you press the two conductive pads together and opens the circuit when you release them.

1. Cut the thick felt into a circle about 2 inches (5 cm) across. Cut out the center to make a ring. (Or leave the felt square and make a frame.) It's OK to cut through the ring to get to the center.

2. With the embroidery floss, sew the ring to one piece of aida cloth with a running stitch, keeping the aida cloth tight.

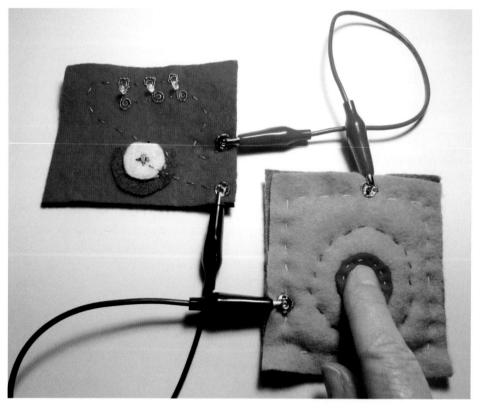

FIGURE 5-38: **A quilted button lights an LED to show that you have pressed it.**

3. To make the first conductive pad, take the conductive thread and fill the center of the ring with Xs. Leave the rest of the thread attached for now.

4. To make the second conductive pad, place the second piece of aida cloth on top of the ring. Line up the edges with the first piece of aida cloth. Trace around the center of the ring. With the conductive thread, fill the circle with Xs. Leave the rest of the thread attached for now.

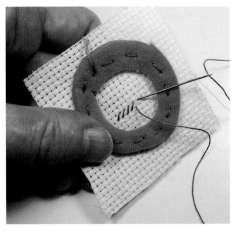

FIGURE 5-39: **Sew a felt ring to a piece of stiff cloth.**

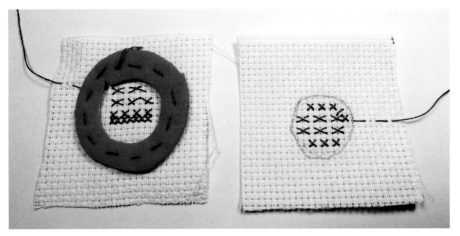

FIGURE 5-40: **Both conductive pads**

5. Put the second piece of aida cloth on top of the ring. Make sure that the two loose pieces of conductive thread are not close enough to touch. With the embroidery floss, sew the two pieces of aida cloth together by making a running stitch around

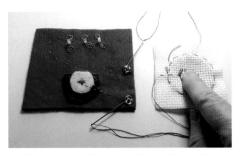

FIGURE 5-41: **Sew the aida cloth together and test out the button.**

the outside of the ring. Pull the top piece of aida cloth tight as you sew. The two conductive pads should not be able to touch each other unless you press on the top.

6. To make a quilted cover for the button, lay the inside of the sensor on the bottom piece of felt. Make sure the loose pieces of conductive thread are laid straight out and that they point in different directions. Next place the batting on the sensor. Cover the batting with the top piece of felt. Pin the pieces together, again making sure the loose pieces of conductive thread are sticking out. (See the quilt project in Chapter 3 for pinning directions.) Use embroidery floss to sew the edges closed with a running stitch or blanket stitch. Avoid catching the conductive thread with your needle. The loose ends of the conductive thread should stick out between the stitches you are making.

7. With the loose ends of the conductive thread, sew on snaps or make thread pads along the edge of the sensor.

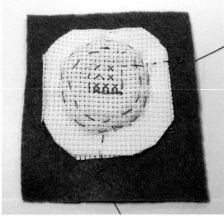

FIGURE 5-42: **The sensor on the bottom piece of felt**

FIGURE 5-43: **The batting**

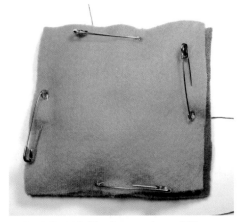

FIGURE 5-44: **Pin the top piece of felt on.**

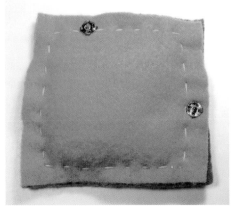

FIGURE 5-45: **Sew on snaps.**

8. With embroidery floss, quilt around the outside of the sensor ring through all the layers to make a puffy "button." You can sew on a circle of felt in a different color to mark it. Test your button to see if it works!

FIGURE 5-46: **The back of the quilted pressure sensor**

II. The Soft Tilt Sensor

FIGURE 5-47: **A tilt sensor closes a circuit when it is off center.**

When you tilt this sensor, a swinging beaded string closes the circuit and makes the lights on the circuit tester light up.

1. Start with a thick or doubled piece of felt for a base. With the conductive thread, sew a snap or pad in one upper corner. Then sew a row of running stitches from there to the middle of the upper edge. Anchor the thread with a few stitches. Insert the needle under a stitch to make a knot. End with the thread on the right side of the fabric (the front of the sensor).

FIGURE 5-48: **Anchor the thread.**

2. To make the beaded string, slide some small beads onto the thread. Then slide on one or more large conductive beads.

3. To keep the beads on the string, wrap the thread around the last bead. Then insert the needle back into the beads and go back the other way. When you get to the top of the string of beads, pull the thread tight. Then take a few stitches in the felt to anchor the string of beads again.

4. Tilt the sensor so the beaded string swings as far to one side as you want it to go. Mark the spot where the last conductive bead touches the felt. Repeat with the other side. This is where your conductive stop will go. When the conductive bead touches one of the conductive stops, it will close the circuit and cause the LED to light up.

5. With a new long piece of conductive thread, sew a row of small conductive beads on one of the marks you made to make the conductive stop. (You can also try a snap or a thread pad.)

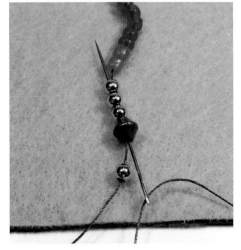

FIGURE 5-49: **Wrap the conductive thread around the last bead on the string. Then insert the needle back into the beads and go back the other way.**

FIGURE 5-50: **The string of beads**

Make sure the conductive bead on the string makes good contact with the beads on the stop when the sensor is tilted toward that side.

6. With the same piece of conductive thread, sew a row of running stitches across the base to the other conductive stop. If the felt is thick enough, poke the needle through the base and sew your stitches on the back (wrong side) of the base. Or to make a row of stitches across the front (right side) of the felt, lightly mark the path of the string as it swings from side to side and make sure your stitches go below that line. With the same piece of conductive thread, make the second conductive stop, the same way you did with the first in Step 5.

FIGURE 5-51: **Sew on beads.**

FIGURE 5-52: **Beads on the left and right both connect to the other snap.**

FIGURE 5-53: **Sewing the stitches across the back of the thick felt prevents the swinging beads from causing a short circuit.**

7. To test your Tilt Sensor, use alligator wires to connect the conductive pads to the Tester.

FIGURE 5-54: **The finished tilt sensor**

III. *The Stroke Sensor*

Conductive "hairs" bend when you stroke them and make contact with other conductive hairs to close a circuit. An ingenious variation on a touch sensor!

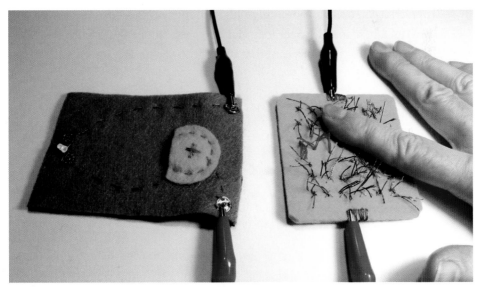

FIGURE 5-55: **Brush the conductive hairs on the stroke sensor to make the LEDs light up.**

1. To make the base for this sensor, take a piece of thick felt or doubled piece of felt and sew a conductive pad along one edge.

2. Using the same piece of thread, make two or three rows of Xs across that end of the base. Leave the middle third clear.

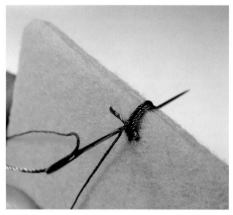

FIGURE 5-56: **Make a conductive pad with the conductive thread.**

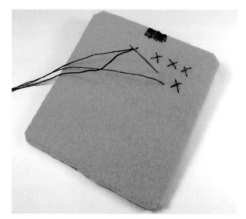

FIGURE 5-57: **Sew Xs across the base.**

3. Anchor this thread (or a new piece) on one of the Xs. To begin making loops, bring the needle up from underneath the felt and poke it back down very close to the first hole. Put your fingertip or a pencil through the loop of thread so you don't accidentally pull it all the way closed. On the wrong side of the felt, insert the needle under a nearby stitch. Tie the thread to the stitch. Repeat this process until that end of the sensor is covered with loops.

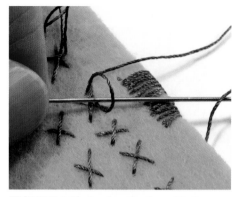

FIGURE 5-58: **Anchor the thread before making a loop.**

FIGURE 5-59: **A loop tied to an X**

4. Cut the loops open so they become hairs. Test the hairs by clipping one alligator wire from the Tester to the connecting pad on the Stroke Sensor. Brush the hairs with the other wire. The LEDs on the Tester should light up. If they don't, check to see that they are connected to the conductive pad.

FIGURE 5-60: **One end covered in loops**

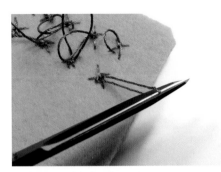

FIGURE 5-61: **Cutting a loop**

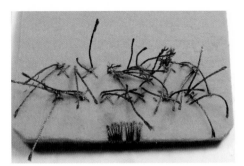

FIGURE 5-62: **Loops cut into hairs**

5. Repeat Steps 1 through 4 with a new piece of conductive thread on the other end of the sensor.

6. Time to fill in the middle of the Stroke Sensor. The trick is to be sure that the hairs on one side make an electrical connection with the other side only when you brush them. Here are some ways to solve this design problem:

⊙ If you didn't leave enough space between the two ends, or if the hairs are too long, they may connect simply by flopping over. Trim

FIGURE 5-63: **Finished Stroke Sensor**

them so they can stand up better. They should connect only when you bend them. Fill the area in the middle with hair made of non-conductive thread.

⊙ If the hairs on the two ends can't reach each other, you will need to make a conductive path to act like a bridge between them. First, sew a few tufts of conductive thread hair in the middle. Be careful not to connect these conductive threads to either end! Make sure this bridge works. Then fill in the rest of the middle with non-conductive thread.

IV. The Spool Knitter Stretch Sensor

The special steel/polyester blend yarn in this sensor lets varying amounts of current through depending on how tightly you stretch it.

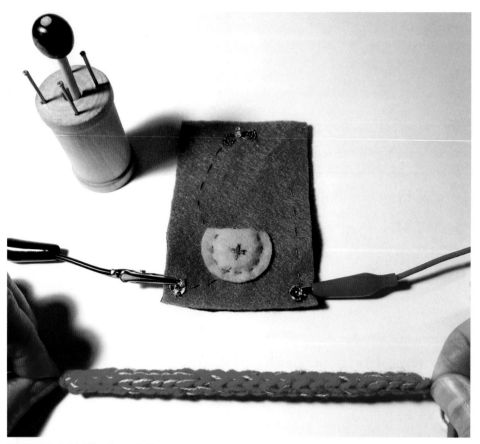

FIGURE 5-64: **The Stretch Sensor**

1. To knit a tube using stretch-sensitive conductive yarn and regular yarn, tie the ends of both yarns together. Insert the knotted ends into the top of the hole in the spool knitter and pull it out through the bottom. Leave a long tail.

2. Hold the two strands of yarn together and use them the same way you would use a single piece of yarn. Cast on and knit as usual. (See Chapter 4 for directions.) Be careful to keep yarn together for every stitch, and keep the tension even. Keep going until you knit a tube about 6 inches (15 cm) long.

3. Cut both yarns, leaving a long tail. Cast off as usual. Pull the knitted tube out of the knitter. Pull the tails of both ends to close them up. If any of the stitches are very loose, try to even out the tension by pulling gently on the loose stitch with your fingers or a crochet hook.

4. To use the stretch sensor, attach it to your Tester with alligator wires. (The bar graph version with three LEDs works very well with this sensor.) Make sure the metal teeth on the alligator clip clamp onto the

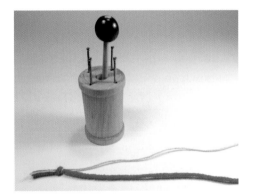

FIGURE 5-65: **Tie the conductive and regular yarn together.**

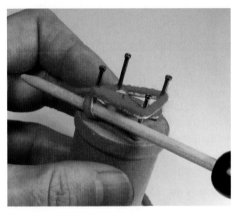

FIGURE 5-66: **Knitting both strands together**

FIGURE 5-67: **Make a tube about 6 inches (15 cm) long.**

conductive yarn. Do the lights get brighter (or dimmer) when you stretch your sensor? (Look closely; the difference may be very slight.) To increase the effect, try making a longer stretch sensor, or think about how you could stretch it tighter (such as wrapping it around your hand or getting a friend to pull on the other end).

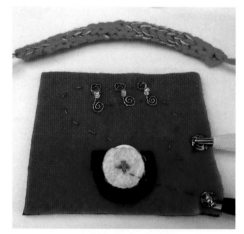 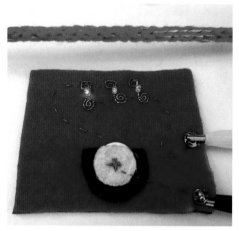

FIGURE 5-68: **One light shines when the Stretch Sensor is relaxed.**

FIGURE 5-69: **More lights shine when the sensor is pulled tight.**

Extensions:

Make a stand-alone soft circuit project by combining one or more of these sensors with a battery and an LED, motor, buzzer, or other load. For example, you can make a patchwork quilt checkerboard with a pressure sensor on/off switch that lights up through "windows" made of lighter-colored fabric. Or use the Stroke Sensor to make a stuffed hedgehog, cactus, caterpillar, or other spiny design that squeaks using a buzzer or shivers using a vibration motor.

Here is one quick stand-alone project: a tilt sensor fob for a zipper pull or key chain. The battery pouch is shaped like a cat (or another animal of your choosing). When the tail swings, it makes the LEDs flash on and off:

1. Start with a base of thick stiffened felt about 3 by 4 inches (8 by 10 cm), or use peel-and-stick felt or an iron-on adhesive sheet to make a doubled layer of regular felt. Cut the bottom in a curve.

2. Take a coin battery and place it on the base, just above the center. Use a pencil to trace around it.

3. Prepare one or more LEDs by bending the wire leads into spirals—round for the positive lead, square for the negative lead. Place the LEDs along the bottom of the base, following the curve, with the negative (square) leads pointing down. Make sure the LEDs are not touching each other. Use conductive thread to sew the negative lead of the first LED to the base. Make a stitch through the base toward the next LED and sew that negative lead on the same way. Repeat with the remaining LEDs.

4. Now poke the needle with the conductive thread through to the back of the base. If the felt is thick enough, sew a row of running stitches from there up to the spot where the battery will go—but don't let the stitches show through on the front of the base. (If the felt is not thick enough, tape or glue the thread to the back.) When you reach the spot for the battery, bring the needle up through the felt to the front of the base. Make a conductive pad with a few stitches, then anchor and cut the thread.

5. Take another piece of conductive thread and sew the positive lead of the first LED onto the base. Anchor and cut it when you are done. Repeat with the other LEDs, making sure the thread and the leads do not touch any of the other LEDs.

6. To make the battery pouch, take another piece of felt and trace around the battery. Then draw a cat or other shape around it. Make sure your design fits on the base, then cut it out along the outline you drew.

7. Sew a tail on the battery pouch using conductive thread and conductive beads. (See Steps 1–3 of the Soft Tilt Sensor project.) Make the tail long enough to touch the positive leads of the LEDs.

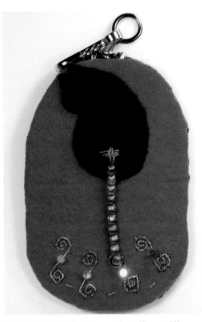

FIGURE 5-70: **The cat's tail acts like a tilt sensor connector to turn the lights on as it swings past.**

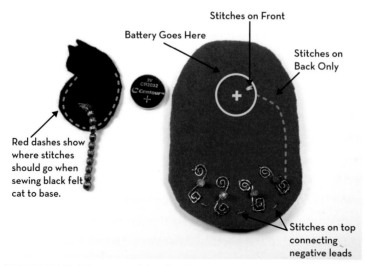

Stitches on Front

Battery Goes Here

Stitches on Back Only

Red dashes show where stitches should go when sewing black felt cat to base.

Stitches on top connecting negative leads

FIGURE 5-71: The parts of the circuit

8. To sew the battery pouch onto the base, use embroidery floss or non-conductive thread. Stitch around the bottom in a "U" shape, like you did with the battery pouch on the Tester. Leave enough room at the top of the pocket to slip the battery in and out. (Black floss was used in the example in Figure 5-72 so it would blend in with the cat.)

9. To hang your fob on a zipper pull or key chain, sew a lanyard snap clip or a split key ring onto the back of the base. Cover up the back by sticking on another piece of felt to avoid short circuits. Then insert the battery, positive side up, into the pocket and your flashing fob is ready to go!

FIGURE 5-72: The back of the fob. Note the circuit trace that connects the negative leads and goes up to the battery pad. The positive leads have separate pads that are not connected to anything else. Nonconductive black thread holds the battery pouch on the base. The clip at the top is also sewn with nonconductive thread.

PROJECT:
WOVEN AUDIO SPEAKER

◉ MATERIALS

1 or more colors of thin cotton yarn (like the kind used for baby clothes) for the warp and weft

Small Styrofoam bowl (or plate)

Conductive thread

3 small beads (that fit on the conductive thread)

Musical greeting card

(Optional) aluminum foil and electrical or other insulating tape

1 or more magnets, the stronger the better, such as a rare earth magnet or a stack of 3–4 ceramic disk magnets (you can also remove and use the magnet that comes in the speaker from the greeting card)

◉ TOOLS

2 needles—one for the yarn (preferably short with a rounded tip), and one for the conductive thread (you will use both at once)

Hot glue gun

Wire clipper and wire stripper (you can use craft scissors, but be careful not to cut all the way through the wire)

2 alligator clip wires

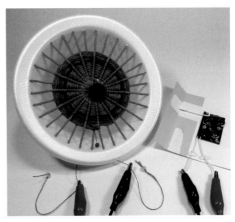

FIGURE 5-73: **Finished speaker playing**

A simple speaker consists of a coil of conductive wire, a paper or fabric diaphragm that vibrates, and a magnet. (To find out how it works, see my book *Musical Inventions*.) For this low-tech speaker, you'll weave a coil from conductive thread and cotton yarn. But instead of a square cardboard loom like the one you made in Chapter 3, this loom is made from a disposable bowl—which also serves as a cone to direct the sound to your ears. All you need to make it play is a magnet!

1. Start by removing the music system from the greeting card. Open up the glued flap inside. Behind it sits the speaker and the circuit board that plays the music. Look for a plastic strip that keeps two metal strips from touching. This is the switch that turns the music on when you open the card. If you need to, cut it or make a substitute from a piece of cardboard so you can turn the music on without the rest of the card. Cut the sound system out of the rest of the card, leaving a small piece of cardboard for it to sit on.

2. Carefully cut the wires that attach the speaker to the circuit board—but leave as much of the wire still attached to the board as you can. Peel the speaker off the card. Carefully strip about ½ inch (1.25 cm) of the insulation off the ends of its wires.

3. You can also take the speaker apart to see how it works. Inside you'll find a coil of thin copper wire and a magnet. You can reuse either or both of these in your design.

FIGURE 5.74: **A musical greeting card**

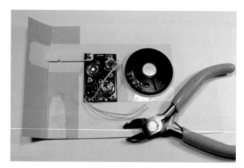

FIGURE 5.75: **Clip the wires connecting the circuit board and the speaker as close to the speaker as possible.**

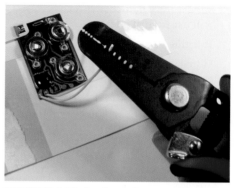

FIGURE 5.76: **A wire stripper makes it easier to remove the insulation from the wires.**

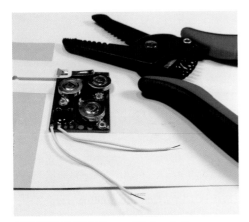

FIGURE 5.77: **Wires with insulation removed from the ends**

FIGURE 5.78: **The card's speaker consists of a plastic cone, a wire coil, and a magnet.**

4. Time to dress your circular loom! (See the Cardboard Loom project in Chapter 3 for the basics.) To start, thread the yarn needle with about 9 feet (3 m) of the warp thread. Tie a knot in the end. Take the needle and poke a hole from the outside to the inside of the Styrofoam bowl, just below the rim. Pull the yarn all the way through,

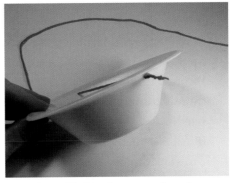

FIGURE 5.79: **The first warp thread**

stopping when you hit the knot. Then carry the needle across the center of the bowl and poke it through the opposite side. Pull the yarn through the hole until it is tight (but not tight enough to bend the bowl).

5. Next, go about ½ inch (1.25 cm) to the right of the second hole and poke the needle back into the bowl, from the outside to the inside. Bring the yarn across the middle of the bowl again, and cross over the first warp thread, making an X. On the opposite side of the bowl, poke the needle out through the side about ½ inch (1.25 cm) to the left of the last thread and pull the yarn through. Continue crisscrossing the bowl with the warp thread in the same way until you come back around to the other side of the first hole. Then tie off the end of the yarn, trying to keep it tight.

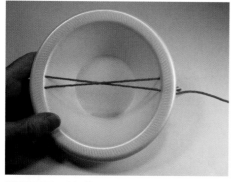

FIGURE 5.80: **Cross the warp threads in the middle of the bowl.**

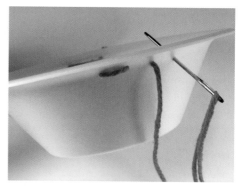

FIGURE 5.81: **Continue adding warp threads all around the bowl.**

6. Cut a piece of conductive thread at least 8 feet (2.4 m) long and thread it on its own needle. Slide one of the beads on, about 10 inches (25 cm) from the end. Insert the needle through and tie a knot around the bead to hold it in place. From underneath the bottom of the bowl, poke the needle up through the foam and through the center of the weaving and pull it through until the bead meets the bottom of the bowl. Tie the conductive thread around the warp thread in the center. Leave the thread on the needle, but put it aside for now. (Stick the needle with the thread on it into a pin cushion, or put your work on a pillow or folded towel that you can stick the needle into so you don't lose it.)

FIGURE 5.82: **Poke the conductive yarn through the bottom of the bowl.**

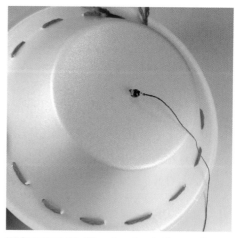

FIGURE 5.83: **Tie a bead onto the conductive thread to keep the long tail from pulling through.**

7. You're ready to start weaving your coil! Cut a long piece of the weft yarn. Thread one end on the yarn needle, and tie the other end over the conductive thread in the center of the warp thread.

FIGURE 5.84: **Tie the conductive thread to the center of the warp threads.**

8. Starting in the center, weave the weft yarn over and under the warp thread. Stop when you get back to the first stitch and set the needle aside.

9. Pick up the conductive thread needle and weave the next row. The stitches should go under and over the warp threads, opposite the yarn stitches. When you get back to the first stitch, switch back to the yarn needle. Continue weaving your coil, alternating one row of yarn and one row of conductive thread. Work slowly and carefully so every row of conductive thread is always insulated by rows of yarn.

FIGURE 5.85: **Tie the weft yarn over the conductive thread.**

FIGURE 5.86: **The first round of weaving with yarn**

FIGURE 5.87: **Alternate rounds of yarn and conductive thread.**

FIGURE 5.88: **Weave the rounds close together.**

10. When you have about 1 foot (30 cm) of conductive thread left, poke the needle through the side of the bowl to bring the tail outside—but first, tie a bead on to keep the thread in place. Tie on another bead on the outside of the bowl. Seal the ends of the thread with a dab of hot glue to keep them from unraveling. Let it dry.

FIGURE 5.89: **Use beads to hold the thread in place.**

11. To use the speaker, connect one of the wires from the greeting card sound system to one of the conductive thread tails with an alligator wire. Do the same with the other wire. Place the bowl on top of a magnet. If you have two or more, place one magnet underneath the bowl and one inside to hold each other in place. You can also place them on either side of the weaving. Start the recorded music by letting the metal strips on the circuit board connect. It will be soft, so get close. Can you hear anything?

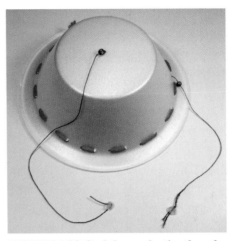

FIGURE 5.90: **Seal the conductive thread ends with glue.**

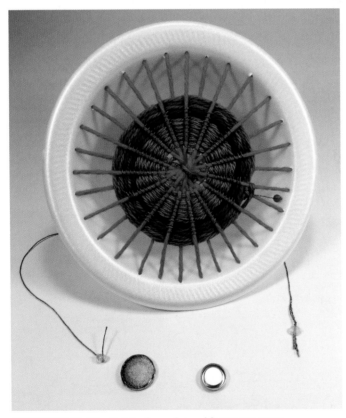

FIGURE 5.91: **The finished speaker with two magnets**

Extensions and variations:

▣ To hear your speaker better, try these suggestions:

  ⊙ Use a stronger magnet, or stack more magnets together.

  ⊙ Place the speaker on a larger container.

  ⊙ Experiment with other kinds of conductive fiber, such as thin copper magnet wire, jewelry wire, or the thin copper ribbon from a pot scrubber. (If using wire that is insulated, rub the coating off the ends with sandpaper.)

▣ To attach the speaker to the card permanently, twist each metal wire around one of the conductive threads. Take a tiny piece of aluminum foil and squeeze it around the twisted wire and thread to hold them in place. Cover the foil with electrical tape for another layer of protection.

▣ To add an audio jack to your speaker so you can plug it into your phone, portable music player, or battery-powered radio, see the Super Simple Speaker project in my book *Musical Inventions*.

▣ Try making a coil using other fiber techniques, such as knitting, crochet, or embroidery.

▣ Add e-textile speakers to a hat, headband, scarf, or ear warmers. Get creative!

INDEX

A

acrylic paint, 5
Adafruit, 163
adhesives, requirement, xix
aluminum foil, 159
amigurumi, 125
arm warmers, 38
atoms, explained, 152
audio speaker, making, 189–196

B

backstitch, xxiii–xxiv
ball of t-shirt yarn
    connecting pieces, 59
    making, 56–59
    materials and tools, 56
basket. *See* coil basket
baste stitch, xxiii
batteries, positive and negative sides,
    153. *See also* soft circuits
batting
    trimming, 109
    using for quilting, 101, 107
black paint, 4
blue pigment, color wheel, 4
brainstorming ideas, xii
Buechley, Leah, 150, 163, 165
bull's-eye, creating for tie dye
    t-shirts, 16
buzzers, 162

C

Cano-Murillo, Kathy, 23

Capillary Action Sun Print, making,
    9–13
cardboard weaving loom
    making, 82–88, 92
    materials and tools, 82
checkerboard, making from quilting
    fabric, 102–111
Chibitronics Circuit Stickers, 163
choker, creation, xiii–xiv
chromatography, 23
circuit board, explained, 155. *See also*
    soft circuits
circuit components, series and
    parallel, 154
circuit measurements, 153
closed circuit, 152–153
coal tar, 3
coaster, making on cardboard loom,
    82–88, 92
cocoa mug. *See* hot cocoa mug/roll-up
    scarf
coil basket
    making, 75–79
    materials and tools, 75
coin batteries, working with, 153
collars or cowls, 40
color and light, 3–4
color wheel, 4–5
colored dyes, 2
colors
    mixing, 4
    relationships, 5
conductive fabrics, 158

conductive materials, 152

conductive pads, making, 159–160

conductive ribbon, 158

conductive thread, working with, xx, 156–158, 168. *See also* thread

copper pot scrubber, 159

cotton fabric, 6

crochet hook, requirement, xviii

crochet hot cocoa mug/roll-up scarf, 125–131

crocheting
    chain (ch), 119
    changing yarn, 123
    decrease (dec), 121–122
    fasten off, 123
    holding yarn and hook, 119
    hot cocoa mug/roll-up scarf, 125–130
    increase (inc), 121
    single crochet (sc), 120
    slip knot, 118
    slip stitch (sl st), 122
    weave in ends, 124
    yarn over, 119

current, resistance, and voltage, 153

cyanotype, 8–9

D

DIFF, 94

dyes, coloring, 2

dyes and paints
    best practices, 6
    minimizing mess, 6
    safety warning, 16
    tips, 5–6

E

electrical circuits, making, 152–155

electrical components, positive and negative ends, 154

electricity, explained, 152

electronic components, connecting, 159–160

embroidery hoop, requirement, xviii

The Empowerment Plan, 93

EMPWR Coat, 93

EPS (Elizabeth Percentage System), 141

e-textiles, xiv, 150
    requirement, xx–xxi

F

fabric, requirement, xviii

fabric adhesive tape, 158

fabric dyes, 6

fabric marker, requirement, xvi

fabrics
    cotton, 6
    preparing for coloring, 6

Fair Isle electronic circuit, 117

felt
    safety warning, 164
    working with, 164

felted sweater hat
    making, 44–47
    materials and tools, 44

felted sweater mittens
    making, 42–43
    materials and tools, 42

felted sweater wearables
    arm warmers, 38
    collars or cowls, 40
    leg warmers/boot toppers, 37–38
    materials and tools, 37
    scarves, 39–40
    shrugs, 40–41

felted wool sweaters, upcycling, 35–36

felting, 34

floral wire, xxi, 159

French knitter, 142

Fried, Limor, 163

fringed t-shirt
  making, 54–55
  materials and tools, 54

G
Gemma board, 163
goals, narrowing down, xii
green pigment, color wheel, 4

H
hand stitches, xxii–xxvi
hats, making from felted sweaters,
  44–47
headband, making from t-shirt yarn,
  60–64
hot cocoa mug/roll-up scarf. *See also*
  scarves
  cocoa and whipped cream
    sections, 127
  handle and mug, 128–129
  materials and tools, 125
  wearing scarf, 130
  whipped cream section, 126

I
ideas, brainstorming and testing, xii
improvements, making, xii
indigo pigment, 2
Instructables.com, xiii, 117
insulators, 152
inventions, sharing, xii
iron and ironing board, requirement,
  xvii

J
Jacquard, Joseph Marie, 81
jewelry wire, xxi, 159

K
knitted tube, using, 147
knitting
  binary aspect, 114

cast off, 136
cast on, 131–132
garter stitch, 134
purl stitch, 133
ribbing, 135
stitch, 132–133
stockinette stitch, 135
tiny toothpick knitting,
  137–139
turning, 134
knitting nancy, 142
knitting sweaters, 116–117
knots, xxvi
KOBAKANT collective, 165–166

L
LEDs (light-emitting diodes). *See also*
  soft circuits
  with leads, xx
  in series and parallel, 154
  testing on batteries, 160–161
  working with, 160–162
leg warmers/boot toppers, 37–38
light and color, 3–4
LilyPad Arduino toolkit, 150, 163
loads
  relationship to circuits, 154
  types, 162
looms, weaving on, 80
Luna, Angela, 93–94

M
magnet wire, xx, 159
malaria, treatment, 3
masking tape, xvi
materials and tools
  adhesives, xix
  ball of t-shirt yarn, 56
  cardboard weaving loom, 82
  coil basket, 75
  crochet hook, xviii
  embroidery hoop, xviii

e-textile tools and
    supplies, xx–xxi
fabric, xviii
fabric marker, xvi
felted sweater hats, 44
felted sweater mittens, 42
felted sweater wearables, 37–41
fringed t-shirt, 54
hot cocoa mug/roll-up scarf, 125
iron and ironing board, xvii
masking tape, xvi
needle threader, xvi–xvii
needles, xvi
notions and fasteners, xix
ombre t-shirt, 22
paints, xix
quilted checkerboard, 102
repurposed and upcycled, xxi
safety pins, xvi
scissors, xv–xvi
seam ripper, xvii
silkscreening, 25
sock creatures, 65–66
soft circuit tester, 167
soft sensors, 173
spool knitter, 140
straight pins, xvi
sun prints, 7
thimble, xvii
thread, xix
tie dye t-shirts, 14
tiny toothpick knitting, 137
t-shirt tote bag, 50
t-shirt yarn knotted headband, 60
upcycling felted wool sweater, 35
wearable shelter, 95
woven audio speaker, 189
yarn, xix
mauve pigment, 3
Maxwell, Rebecca Angel, 125
Micro:bit mini-computer, 163
microcontrollers, 163

mittens, making from felted
    sweaters, 42–43
motionboard, 162
motors, 162
musical greeting card, using, 190

N
needle arts tips, 115
needle threader, requirement, xvi–xvii
needles
    requirement, xvi
    threading, xxii
neutrons, explained, 152
Newton, Isaac, 3
no-sew projects
    felted sweater wearables, 37–41
    t-shirt tote bag, 50–53
notions and fasteners, requirement, xix

O
ombre t-shirt, 22–24. *See also* tie dye
    t-shirts; t-shirts
        making, 23–24
        materials and overview, 22–23
online resources, xx
orange pigments, color wheel, 4
overcast stitch, xxiv
overhand knot, xxvi

P
paints, acrylics, xix
paints and dyes
    best practices, 6
    minimizing mess, 6
    safety warning, 16
    tips, 5–6
Perkin, William Henry, 3
Perner-Wilson, Hannah, 165–166
photos, turning in Capillary Action
    Sun Print, 12
Pi Shawl, 141
pigments. *See* colors

pliers, xxi
primary colors, 4–5
projects
  ball of t-shirt yarn, 56–59
  coil basket, 75–79
  crochet hot cocoa mug/roll-up
    scarf, 125–131
  felted sweater hat, 44–47
  felted sweater mittens, 42–43
  felted wool sweater
    upcycling, 35–36
  fringed t-shirt, 54–55
  no-sew felted sweater
    wearables, 37–41
  no-sew t-shirt tote bag, 50–53
  ombre t-shirt, 22–24
  quilted checkerboard, 102–111
  silkscreening, 25–30
  sock creature, 66–72
  soft circuit tester, 167–172
  soft sensors, 173–188
  spool knitter, 140–147
  sun prints, 7–13
  tie dye t-shirts, 14–21
  toothpick knitting, 137–139
  t-shirt yarn knotted
    headband, 60–64
  wearable shelter, 95–100
  weaving loom, 82–88
  woven audio speaker, 189–196
protons, explained, 152
ProtoSnap LilyPad Development
  Board, 163
prototype, creating, xii
purple pigments, color wheel, 4

Q
Qi, Jie, 163
quilted checkerboard
  hemming, 110–111
  lining up rows, 105
  materials and tools, 102
  pinning together layers, 107
  placing batting, 107
  sewing strips together, 103
  stitching seams, 108
  trimming batting, 109
  undoing mistakes, 106
quilted pressure sensor, making,
  174–177
quilting overview, 101

R
Ravelry knitting group, 117
red pigment, 2, 4
research, doing, xii
resistance, current, and voltage, 153
resistor steel polyester yarn, xx. *See
  also* yarn
Riddell, Harriet, 112
ripstop nylon, using for wearable
  shelter, 93, 96–100
running stitch, xxiii

S
safety pins, requirement, xvi
safety warnings
  felt, 164
  silkscreening, 26
  tie dye t-shirts, 15
satin stitch, xxv
Satomi, Mika, 165
scarves. *See also* hot cocoa mug/
  roll-up scarf
    designing, 131
    making from felted
      sweaters, 39–40
scissors, requirement, xv–xvi
Scott, Veronika, 93
seam ripper, requirement, xvii
Seay, Jesse, 116–117
secondary colors, 4–5
sensor, relationship to electricity, 155.
  See also soft sensors

sewing equipment, xv
sewing machine, using, xxvii–xxviii
sewing techniques, hand stitches, xxii–xxvi
sharing inventions and stories, xii
shelter project, 95–100
short circuit, explained, 155
shrugs, 40–41
silkscreening
    designing, 26
    making, 27
    materials and tools, 25
    overview, 25–26
    printing, 28
    safety warning, 26
    saving stencil, 29
    test piece, 30
skein of yarn, 115
slip stitch, xxv
"smart clothing," 150
sock creature, 66–72
sock creatures
    making, 66–72
    materials and tools, 65–66
soft circuit tester. See also batteries
    materials and tools, 167
    sewing, 167–172
soft circuits. See also batteries; circuit
    board; LEDs (light-emitting
    diodes)
    conductive pads, 159–160
    conductive thread, 156–158
    materials for traces, 158–159
    power for, 153
soft sensors. See also sensor
    materials and tools, 173
    quilted pressure sensor, 174–177
    soft tilt sensor, 178–181
    spool knitter stretch
        sensor, 184–188
    stroke sensor, 181–184
soft tilt sensor, making, 178–181

solder, using with Fair Isle
    sweater, 117
speaker, making, 189–196
spectrum of light waves, 3–4
spool knitter
    making, 142–147
    materials and tools, 140
    stretch sensor, 184–188
square knot, xxvi
Stern, Becky, xiii–xiv
stitches
    beginning, xxii
    evenness, xxii
stories, sharing, xii
straight pins, requirement, xvi
stroke sensor, making, 181–184
sun prints
    capillary action, 9
    making, 9–13
    materials and tools, 7
    overview, 8–9
Swap-O-Rama-Rama, 32
sweaters, knitting, 116–117. See also
    wool sweaters
switch, relationship to electricity, 155

T
Teknikio, 162
testing ideas, xii
textile medium, 5
thimble, requirement, xvii
thread. See also conductive thread
    cutting, xxii
    knotting, xxii
    pulling, xxii
    requirement, xix
tie dye t-shirts. See also ombre t-shirt;
    t-shirts
    blending colors, 19
    bull's-eye, 16
    bull's-eye pattern, 21
    handling dye, 19

materials and tools, 14
removing excess dye, 19
safety warning, 15
spiral, 18
stripes, 17
Tulip One-Step Tie Dye, 19
V stripe pattern, 21
tiny toothpick knitting
materials and tools, 137
project, 137–139
tools and materials
adhesives, xix
ball of t-shirt yarn, 56
cardboard weaving loom, 82
coil basket, 75
crochet hook, xviii
embroidery hoop, xviii
e-textile tools and
supplies, xx–xxi
fabric, xviii
fabric marker, xvi
felted sweater hats, 44
felted sweater mittens, 42
felted sweater wearables, 37–41
fringed t-shirt, 54
hot cocoa mug/roll-up scarf, 125
iron and ironing board, xvii
masking tape, xvi
needle threader, xvi–xvii
needles, xvi
notions and fasteners, xix
ombre t-shirt, 22
paints, xix
quilted checkerboard, 102
repurposed and upcycled, xxi
safety pins, xvi
scissors, xv–xvi
seam ripper, xvii
silkscreening, 25
sock creatures, 65–66
soft circuit tester, 167
soft sensors, 173

spool knitter, 140
straight pins, xvi
sun prints, 7
thimble, xvii
thread, xix
tie dye t-shirts, 14
tiny toothpick knitting, 137
t-shirt tote bag, 50
t-shirt yarn knotted headband, 60
upcycling felted wool sweater, 35
wearable shelter, 95
woven audio speaker, 189
yarn, xix
toothpick knitting, 137–139
traces
conductive materials, 158–159
explained, 155
Tremayne, Wendy, 32
t-shirt tote bag
making, 50–53
materials and tools, 50
t-shirt yarn knotted headband
making, 60–64
materials and tools, 60
t-shirts. *See also* ombre t-shirt; tie dye
t-shirts
ball of yarn, 56–58
with fringes, 54–55
popularity, 48
upcycling, 48–49
Tulip One-Step Tie Dye, 19

U
upcycling
explained, 32
felted wool sweaters, 35–36
t-shirt tips, 49
wool sweaters, 33–34

V
variable resistors, 155
vibration motors, 162

visible spectrum, 3–4
voltage, resistance, and current, 153

W
warp and weft, 80
wearable shelter
    making, 96–100
    materials and tools, 95
weaving
    on looms, 80
    tips, 83
    tricks and shortcuts, 89–91
weaving loom, making from
    cardboard, 82–88
whip stitch, xxiv
white light, 4
wire cutter, xxi
wire stripper, xx
wool sweaters, upcycling, 33–36. *See also* sweaters

woven audio speaker
    beads and thread, 194
    extensions and variations, 196
    finishing with magnets, 195
    materials and tools, 189
    using musical greeting card, 190
    warp threads, 191–193

Y
yarn. *See also* resistor steel polyester
    yarn
    hiding tail, xxiii
    pulling, xxii
    requirement, xix
    skein, 115
    using for coil basket, 76–79
yellow pigments, color wheel, 4

Z
Zimmerman, Elizabeth, 141

Examine Fascinating Ways to Make Robots!

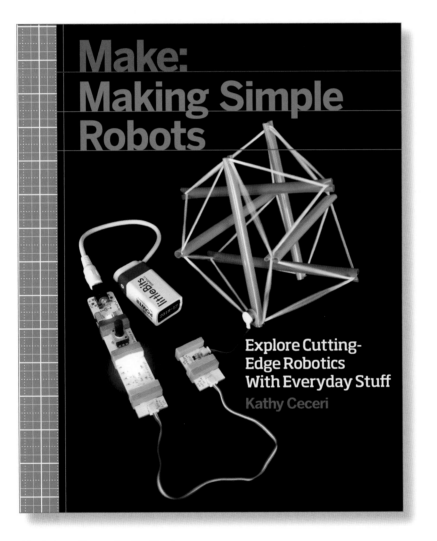

Make:
Making Simple Robots

Explore Cutting-Edge Robotics With Everyday Stuff

Kathy Ceceri

Making Simple Robots

Exploring Cutting-Edge Robotics with Everyday Stuff

ISBN: 9781457183638 | US $24.99

Make:

makezine.com

DISCOVER NEW WAYS TO MAKE YUMMY SNACKS!

Make:
Edible Inventions
Cooking Hacks and Yummy Recipes You Can Build, Mix, Bake, and Grow

Kathy Ceceri

Edible Inventions

Cooking Hacks and Yummy Recipes You Can
Build, Mix, Bake, and Grow

ISBN: 9781680452099 | US $19.99

Make:
makezine.com

Explore the Science of Sound with Your Own Instruments!

Musical Inventions

DIY Instruments to Toot, Tap, Crank, Strum, Pluck, and Switch On

ISBN: 9781680452334 | US $24.99

makezine.com

Learn How to Make Amazing Things with Paper!

Make: Paper Inventions
Machines That Move, Drawings That Light Up, and Wearables and Structures You Can Cut, Fold, and Roll

ISBN: 9781457187520 | US $19.99

Make:
makezine.com